THE OPTICS OF LIFE

THE OPTICS OF LIFE

A Biologist's Guide to Light in Nature

Sönke Johnsen

PRINCETON UNIVERSITY PRESS
PRINCETON AND OXFORD

Copyright © 2012 by Princeton University Press

Published by Princeton University Press, 41 William Street, Princeton, New Jersey 08540

In the United Kingdom: Princeton University Press, 6 Oxford Street, Woodstock, Oxfordshire OX20 1TW

press.princeton.edu

Jacket Photo: An undescribed species of *Hydromedusa* in the genus *Tetrorchis* shows various optical effects, including transparency, absorption by pigments, and iridescence caused by coherent scattering. Courtesy of Steven Haddock.

Library of Congress Cataloging-in-Publication Data

Johnsen, Sönke.
    The optics of life : a biologist's guide to light in nature / Sonke Johnsen.
        p. cm.
    Includes index.
        ISBN 978-0-691-13990-6 (hardback) — ISBN 978-0-691-13991-3 (paperback)   1. Photobiology.   2. Physiological optics.   3. Polarization (Light)   I. Title.
    QH515.J64 2011
    571.4'55—dc23   2011021796

British Library Cataloging-in-Publication Data is available

This book has been composed in Garamond Premier Pro

Printed on acid-free paper. ∞

Printed in the United States of America

10   9   8   7   6   5   4

For my parents, my first and best teachers

CONTENTS

ACKNOWLEDGMENTS

I'm not sure why I wrote this book. I do enjoy writing, but this was a lot of work, so there must have been a reason. When I pitched the project to my editor, I told her that it would fill a niche, but I've never been one to lose sleep over unfilled holes. I also told her that optics was important to biology. It is, but so what? My colleague Steve Vogel told me once that writing books is wonderful because it transforms you from a competitor into an enabler. I do hope this book helps people use optics in their research, but honestly I still feel competitive. Maybe I just want people to stop me in the hall and say, "Nice book!" I'd be lying if I said this didn't matter. I'm shallow, and flattery goes a long way with me.

There's more though. While not conventionally religious, I am often overcome by this world—it's like being given a prize over and over. The most remarkable part to me is that we are able to appreciate and at least partially understand it. Being a biologist, I can mumble about scientific curiosity being an epiphenomenon of natural selection for cooperative hunting, foraging, individual recognition, and so on, but that doesn't make it any less incredible. As the physicist Isidor Rabi said when the muon was discovered, "Who ordered that?" However we acquired this ability to appreciate and understand the world, it would be rude to waste it. So I wrote this book to share this feeling, this amazement at what is all around us.

I got a lot of help. Laurie McNeil, Craig Bohren, and my father, Rainer Johnsen, patiently went over some of the finer physical points with me. Craig, in particular, went far beyond the call of duty, sending wonderfully detailed answers to so many of my questions. Frankly, his books are better than mine; read them if you can understand the math. Laurie McNeil, Eric Warrant, John Endler, Tamara Frank, Marianne Moore, and members of my lab all read parts of this book, helping with language, ideas, and making sure that biologists would understand and appreciate it. Alison Sweeney, Andrij Horodysky, Andrew Smith, my father, and two anonymous reviewers read and edited the whole thing. Many colleagues allowed me to use their figures and photos. A week at Friday Harbor's Helen Whiteley Center finally broke six months of writer's block. Finally, Alison Kalett, Marsha Kunin, Stefani

Wexler, and Karen Carter at Princeton Press turned my draft into a real book.

I'd also like to thank my shipmates over the years, Tammy, Edie, Erika, Steve, Alison, Brad, Justin, Jamie, Dan, and so many others—crazy pirates, all of them—for sharing the excitement and fun of life at sea. Finally, I thank my wife, Lynn, for being there and reminding me—every time I come home excited about some new result—to tell her why she should care, and my daughter, Zoe, for reminding me that "Dad's just a dad."

THE OPTICS OF LIFE

CHAPTER ONE

Introduction

> In the right light, at the right time, everything is extraordinary.
> — AARON ROSE (quoted in *Live in the Light: A Journal of*
>   *Self-Enlightenment*, Mary Engelbreit)

Of all the remarkable substances of our experience—rain, leaves, baby toes—light is perhaps the most miraculous. Essentially indefinable, it is the ultimate food for our planet's life and allows us to perceive the world in nearly magical detail and diversity. Via warmth, vision, and photosynthesis, and its darker aspects such as radiation damage, light interacts fundamentally with nearly all forms of life. Only certain subterranean species may be free from its influence.

Despite this, light remains relatively unstudied by biologists. In my own field of oceanography, we have instruments known as "CTDs" that measure salinity and temperature as a function of depth. These devices are ubiquitous, and the characterization of a body of water is considered incomplete without the data they provide. However, even though light is known to fundamentally affect the distribution, ecology, and behavior of marine organisms, it is seldom measured, despite the availability of economical light-meter attachments made specifically for this instrument. Oceanography is a field known for the tight connections it provides between biology and physics; in other biological fields light measurement is rare. Even worse, many light measurements are taken incorrectly. It is no fun to tell a colleague that, because they didn't put a two-inch cardboard tube around their detector, the data they collected over the last three years is unsalvageable.

In my opinion, the relative lack of optics in biology is primarily the result of a few factors. First, biologists receive very little training in the subject. What they do get is usually confined to the electromagnetism portion of an introductory physics course that derives Maxwell's equations and Coulomb's Law but gives little practical advice about working with light. While there are some good laboratory courses in optics, they're generally populated by physics majors. Second, no other field uses such an arcane and confusing collection of units. Sorting absorbance from absorption, and irradiance from radiance is hard enough without having to do it using nits, candles, and foot-lamberts. Third, the equipment needed is generally geared toward physicists, advertised in their journals, and using their units and terminology. This equipment, while smaller and cheaper than it used to be, is still relatively expensive and fussy compared to, for example, a lab balance.

Finally, and perhaps most importantly, there are few good books. Yes, there are excellent books on optical theory and instrumentation, but they assume a graduate degree in physics or engineering. The three exceptions, *Clouds in a Glass of Beer* and *What Light Through Yonder Window Breaks?* by Craig Bohren, and *QED* by Richard Feynman are wonderful books, but give little practical advice. There are a few excellent books on the optics of vision, my favorite being *Animal Eyes* by Michael Land and Dan-Eric Nilsson, but they do not cover other biologically important aspects of light such as scattering, emission, and absorption.

Therefore, the few biologists working with optics either came in with a physics background (and, like myself, had a lot to learn about biology) or were trained by the even smaller number of biologists familiar with the subject. This has led to a bottleneck where there are many more interesting bio-optical problems than there are people able to work on them. One of my favorite activities as a child was to take apart small natural dams in creeks. It is my hope that this book, by providing the basics necessary to measure and use light in biological research, will breach this bottleneck and lead to a flood of new results and insights.

WHAT IS LIGHT?

Optics is about light, so perhaps we should start with what light is. I have no idea. I have thought about light since I was five years old and am no closer to understanding its fundamental nature. I am in good company though. Even Richard Feynman, one of the creators of the theory of how light and matter interact and widely acknowledged as one of the best explainers of physics, said that light cannot be understood. We have mathematical formalisms that let us predict what light will do to a precision of more than twenty signifi-

Figure 1.1: Darth Vader meets Golden Snitch. *Scientific American's* depiction of a photon in the 1970s and 1980s.

cant figures, but no one has come up with a description of light that makes sense. It is unlikely that anyone ever will.

The root of the problem lies in what is called wave-particle duality, which is usually described as "light sometimes behaves like a particle and sometimes behaves like a wave." In particle language, a beam of light is a stream of photons—small massless particles with energy and momentum that travel at high speeds. In wave language, a beam of light is a series of waves of changing electric and magnetic field strength that have phase, amplitude, and wavelength, and—like photons—travel at high speeds. To intuitively describe what light does, you have to jump back and forth between these two interpretations. This isn't easy. It's hard enough to think of something as being both an apple and an orange, let alone both a microscopic ball and a diaphanous wave that extends throughout space. Combining the two images does no good at all. For example, when I was young, the magazine *Scientific American* depicted photons as little spheres with wave-shaped wings, a ridiculous image that has stuck with me for almost forty years (figure 1.1).

People, especially physical scientists, seem to have an innate intolerance of ambiguity, so many physicists have chosen sides. Some, like Lamb (1995) castigate others for using the photon concept. Others, including Feynman (1985), have stated that light is a particle, end of story. Because wave-particle duality is not limited to light, but is a property of all particles, this has become a fundamental argument about how to describe our universe. Read enough about it, and your head will start to itch.

Maybe light honestly is one or the other, but I prefer a practical approach. In my opinion, sometimes the results of experiments with light are more easily predicted or intuited using the mathematics and metaphors appropriate to waves; and sometimes the results are better explained using the mathematics and metaphors appropriate to particles. I prefer to think about the emission and absorption of light in photon language, imagining photons as little hyperactive balls flying out of light sources and being sucked up by matter. I think of polarization and interference in wave language. These two phenomena can be described using particle language, but it's a bit like using political science to teach someone how to fry a chicken. Scattering is a gray area for me, and I jump back and forth between particle and wave language when thinking about it.

This indecision on my part gets complicated when dealing with quantum mechanics, a topic that I'll visit in the last chapter. However, except for some of the more complicated aspects of photochemistry, the truly weird parts of quantum mechanics are not relevant to biology. In fact, some would argue that, because the uncertainty principle, nonlocality, and other aspects of quantum weirdness play no role in the everyday life of humans, there has been no natural selection for a commonsense understanding of them. So, until we meet a species in which quantum weirdness is experienced directly, and that could possibly explain it to us, I think we're left learning to be comfortable with the fact that some aspects of nature are non-intuitive.

However, while the non-intuitive nature of light can be unsatisfying, it doesn't affect our ability to predict events. In other words, as long as you do your measurements and math correctly, you can think of light as little purple buffaloes and it won't matter. After all, we don't really understand the fundamental nature of anything, but manage just fine.

What This Book Is

The remainder of this book is divided into four major sections. The first short section, comprising only the second chapter, attempts to clear up the mess of units that has turned many people away from optics. The next and largest section, running from chapters 3 through 8, describes the various things that light can do, starting with what Craig Bohren calls the birth, life, and death of a photon (emission, scattering, and absorption) and concluding with fluorescence and polarization. The goal of all these chapters is to provide as unified a view of these processes as possible. So, for example, refraction and reflection are treated as special cases of scattering. Biological examples are given to illustrate various points, but the focus is on explaining these fundamental processes.

The third section, found in chapter 9, is a guide to measuring light. This chapter is meant to be practical and detailed. While not a complete manual for producing publishable measurements and models, it should give you the general lay of the land, and most importantly, identify the major pitfalls.

As mentioned above, the book concludes with a brief chapter on quantum mechanics. While by no means comprehensive, it provides a flavor of the fundamental weirdness of light and all matter. The chapter focuses on two central experiments, the interference of light passing through two slits and the effect of measurement on pairs of entangled photons.

This book is short. I did not want to write an encyclopedia, perhaps because I do not enjoy reading them. Because it is short, I wrote it to be read in its entirety. Later chapters often build on material discussed earlier. While the descriptions in this book are brief and (hopefully!) accessible, they have been vetted for accuracy by a number of experts in optics. I'm sure I still screwed up somewhere, but hope it's nothing serious.

WHAT THIS BOOK IS NOT

What this book is not is vastly larger than what it is. Most importantly, this is not a book about vision or visual ecology. As I mentioned, several excellent authored and edited volumes on human and animal vision exist and can be found in the bibliography. While the book contains some examples drawn from vision research, it is far from comprehensive.

This is also not a book about quantum theory. Again, there are many good books on quantum weirdness, far more in fact than there are on vision. While the last chapter gives a sense of this non-intuitive subject, the bulk of this book deals with classical optics.

This book is also not a comprehensive account of optics or the history of the field. There are massive tomes covering all optical subjects and equally massive and conflicting accounts of the history of the field. It has been said that the person credited with a scientific law is almost certainly not the one who discovered it. Gustav Mie was actually the last person to discover the scattering that bears his name, neither Lambert nor Beer are truly responsible for the Lambert-Beer law of attenuation, and Snell of Snell's Law has had his name spelled wrong for a couple hundred years now (it's actually Snel)

(Bohren and Clothiaux, 2006). I will do my best to give credit where credit is due, but not at the expense of brevity or my central purpose.

Finally, this is not a book for experts. My goal is to briefly and accurately cover what 90 percent of biologists need to know to use optics in their research. As much as possible, complex equations and second-order effects have been omitted. Therefore, if you want to know the difference between Fresnel and Fraunhofer diffraction, what a surface plasmon is, or to delve into the subtleties of nonlinear optics, please consult the excellent texts suggested in the bibliography. I have however, included a number of appendixes that include formulas and constants useful for researchers at any level.

In summary, this book was written to give biologists a brief and clean introduction to this fascinating field. If it does this and encourages you to explore more deeply, via the many excellent and more specialized texts and—more importantly—via your own experience, I'll be satisfied.

FURTHER READING

My favorite popular books on optics are Craig Bohren's *Clouds in a Glass of Beer* and *What Light Through Yonder Window Breaks?*. These two books, mostly consisting of articles from the journal *Weatherwise*, are down-to-earth explanations of various topics in optics and thermodynamics. Never one to accept conventional wisdom, Bohren is insightful, empirical, and fun to read.

If you're up for the math, Bohren wrote another book with Eugene Clothiaux that I think is the best optics text ever written. Despite the rather specific title *Fundamentals of Atmospheric Radiation*, it covers all the essential aspects of optics. It is also oddly (and accidentally) convergent on the book you are holding, possibly because atmospheric scientists have a similar desire to make optics accessible to a larger world. Unfortunately for biologists, it requires a fair bit of mathematical and physical knowledge to fully appreciate. However, you can still get a lot from reading the summaries and conclusions. It is also one of the few physics texts that is actually funny.

Richard Feynman's *QED*, a highly revised transcript of four public lectures on optics and particle physics, should be read by every scientist. It goes from mirrors to quantum physics in 150 simple pages. Even if you don't ever work with light, the book is worth reading to see how much can be derived from so little.

By far the best book on the optics of eyes is *Animal Eyes* by Mike Land and Dan-Eric Nilsson. It manages to explain the diverse eye architectures of the animal kingdom via a few basic principles.

Finally, if you can afford (and can lift!) it, the *Handbook of Optics*, published by the Optical Society of America and edited by Michael Bass is an amazing resource. The 1995 second edition comes in two volumes, its eighty-three comprehensive chapters covering just about everything you would ever want to know, from basic geometric optics to thin-film coatings to optical oceanography to making holograms. There is even a sixty-three-page chapter on making things black. The third edition just came out in 2009 and comprises five volumes and weighs twenty pounds. I can't imagine what else they found to add.

CHAPTER TWO

Units and Geometry

Once upon a time in the land of Nits, dwelt a king named
Troland with his beautiful daughter Candela. And so on.
—CRAIG BOHREN and EUGENE CLOTHIAUX (from
Fundamentals of Atmospheric Radiation)

The number one complaint I get from people who want to learn optics is about units. Although a photon (or light wave) has only three properties—frequency, wavelength, and polarization—the long history of light measurement and its association with human vision has left a trail of perplexing and comical units. Only in optics do people still publish papers using units like stilbs, nits, candelas, trolands, and my personal favorite, foot-lamberts. Unless you work in human visual psychophysics, where these units are entrenched, my advice is simple. If you know what these units are, forget them. If they are unfamiliar to you, never learn them. Instead, stick to the few concepts and units described in this chapter.

PHOTONS VERSUS WATTS

The first issue that most optical biologists confront is whether to measure light in terms of photons or watts. As I mentioned in the first chapter, light can be understood as both a stream of particles and a collection of electromagnetic waves. I'll prove this by shamelessly mixing the two languages right now.

An electromagnetic wave consists of rapidly varying electric and magnetic fields that propagate through space. It has a frequency (usually denoted by the Greek symbol v ["nu"]) and an energy density that is proportional to the square of its electric field. However, this energy can only be imparted to objects in discrete localized units, which we call photons or quanta (whether light actually travels through space in these same discrete units is unanswerable and will keep you up at night). The amount of energy that a photon can impart is proportional to its frequency, though it's hard to think of a particle having a frequency.

Despite this simple relationship between energy and frequency, most people prefer to describe light via wavelength (denoted by λ: "lambda"), possibly because we are more comfortable with lengths than frequencies. The conversion from frequencies to wavelengths is simple: for any wave/photon, wavelength times frequency equals the speed of the wave/photon. There are subtleties to this that we will discuss further both later in this chapter and in chapter 5. For now though, let's say that wavelength equals the speed of light divided by the frequency (i.e., $\lambda = c/v$). This means that the energy of a photon is inversely proportional to its wavelength. You can see that our love for lengths has already made things more complicated.

The electromagnetic spectrum has wavelengths that cover at least fourteen orders of magnitude (figure 2.1), but visible light has wavelengths that only run from about 400 nm to 700 nm. Given the inverse relationship between wavelength and energy, the bluest photons we can see ($\lambda \cong 400$ nm) will have nearly twice the energy as the reddest photons ($\lambda \cong 700$ nm). This can be a problem. Most light meters are calibrated in power units, usually watts (W). Suppose you point such a light meter at a beam of light and get a measurement of 5 W. Can you use this measurement to determine how many photons there were? Unfortunately you cannot, unless you know the wavelength distribution of the light. For example 5 W of 400 nm blue light has only about half the number of photons as 5 W of 700 nm red light. In other words, unless you know how much light there is at each wavelength, you cannot convert from energy units to photons. As I will discuss in chapter 9, it is always best to measure the spectrum of the light (as opposed to some lumped measure), but this is not always possible. So it is important to think about which units are more appropriate for your research, watts or photons.

As I said, most light meters are calibrated for watts. However, it is almost always more appropriate to use photons for biological questions. Biological processes, such as vision and photosynthesis, usually either absorb a photon or don't. The energy of the photon may influence the *probability*

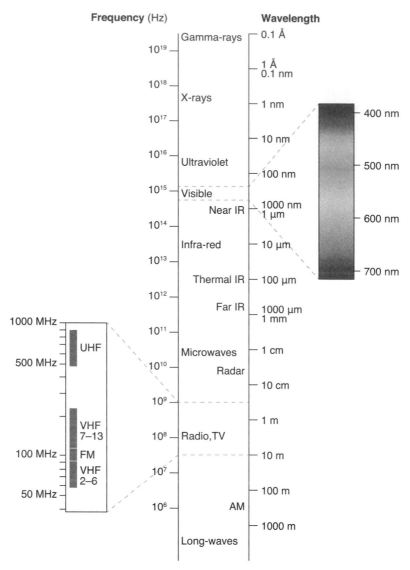

Figure 2.1: The electromagnetic spectrum.

that the photon gets absorbed, but it usually does not affect the result after absorption. For example, the green-sensitive cone cells in our retinas are much more likely to absorb a 550 nm "green" photon than a 650 nm "red" one. However, once absorbed, the wavelength of a photon has no effect on the probability that the photoreceptor will record an electrical event (known as a "photon bump") or on the size of that event. Thus, photoreceptors are often referred to as photon counters. The same is true of photosynthesis, though this gets complicated by the interaction between the chlorophyll in the photosynthetic reaction center and the accessory pigments in the light harvesting complex. Damage from ultraviolet radiation does depend on the energy of the photon absorbed, but the relationship is highly nonlinear. One W/m^2 of 550 nm "green" light might warm your skin. One W/m^2 of 300 nm UVB radiation will severely burn you in a few minutes. The only situation where watts may be more appropriate than photons is the one in which you are interested in simple thermal heating— for example determining how much energy is being absorbed by the skin of a lizard on a hot day.

It is important to remember that watts are not units of energy but units of power (energy per time), so the correct corresponding photon units are not photons alone but photons per second. In most cases, one cares about how many photons arrive over a period of time, not how many photons overall. For example, our perception of the brightness of an explosion depends on how many photons reach our eye per second, not how many reach our eyes over the course of the whole explosion. Also, the density of photons—how many strike a given area—is usually more useful than their total number. This means that you will usually be measuring light in $photons/s/cm^2$. For reasons unknown to me, square centimeters are more often used than square meters, so I will stick with them to avoid confusion. In photosynthesis or other situations where light is supplying the energy for a chemical reaction, the number of photons is often given in moles. A mole of photons is called an "Einstein." This seemed cumbersome to me until a photochemist pointed out that this simplifies the analysis of photochemical reactions since the other reactants are generally given in moles. Because a mole of photons striking a square centimeter every second is intensely bright, values are typically given in microEinsteins.

As I mentioned, it is better to measure the amount of light at each wavelength, rather than all the light over some large wavelength range. If you do this, it is critical to remember that a spectrum is actually a histogram of how many photons fall within each wavelength interval. This sounds obvious, but has serious implications that we will discuss later in this chapter. For now, just accept that a spectral measurement has to include the wavelength interval. Thus, the final units to be used for spectral measurements are photons/s/

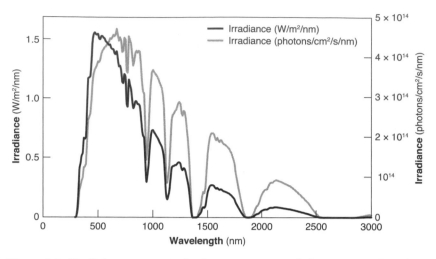

Figure 2.2: Daylight spectrum in both energy units and photon units. Based on the standard reference spectra measured by the American Society for Testing and Materials.

cm²/nm (or watts/cm²/nm) where the "/nm" indicates that you are binning by nanometer-sized wavelength intervals.

If you need to convert a spectrum from watts to photons, simply multiply the value in watts (for a given wavelength) by the wavelength in nanometers and then by 5.05×10^{15}. This works because the energy of a photon is equal to hc/λ (h is Planck's constant and c is the speed of light), so to go from energy to photons you need to divide by hc/λ, which is the same as multiplying by λ/hc. Most people like to use nanometers for the wavelengths of visible light, so $1/hc$ in those units is 5.05×10^{15}.

Since the ratio of watts to photons depends on the wavelength of light, you must know the spectrum of the light you want to convert. If someone hands you a value of 10 watt/cm² that was measured using a detector that lumped light over a large range of wavelengths (for example, the PAR detectors used in oceanography that measure everything from 400 nm to 700 nm) you cannot convert to photons unless you can guess the underlying spectrum. Sounds simple, but this fact has led to many a heartbreak. Spectra in photons/s can look quite different from spectra in watts (figure 2.2), so it is important to know which units you are working in.

Frequency versus Wavelength

As I said earlier, the product of wavelength and frequency equals the speed of light. There are two complications to this. First, there are multiple definitions of the speed of light. We'll discuss this further in chapter 5. For now, speed means the phase velocity, which is the speed that the wave crests of the changing electric and magnetic fields travel. The second complication is that the phase velocity depends on the medium through which the light is traveling. More specifically, the product of λ and v equals the speed of light in a vacuum (c) divided by the refractive index of the medium (n). In other words, $\lambda v = c/n$.

So, suppose that a beam of sunlight goes from air ($n \cong 1$) into the ocean ($n \cong 1.33$). The index goes up, so the phase velocity drops by a factor of 1.33. Since this is the product of frequency and wavelength, one of the two (or both) has to also drop. It turns out that frequency stays the same and wavelength drops. In this case, a "green" 550 nm photon actually has a wavelength of 414 nm in the ocean. So frequency seems to be more fundamental than wavelength. Also, remember the energy of a photon is proportional to frequency, but not to wavelength (the formula for converting watts to photons in the previous section only works if you use the wavelength the light would have in a vacuum). This is important, because in many processes, such as absorption, it is the energy of the photon that matters, not its wavelength. For example, even though the wavelength of a "green" photon inside our eye depends on whether the eye is full of water or air, our perception of it doesn't change, because absorption of light by photoreceptors depends on the energy of the photons, which is related to the unchanging frequency.

As I mentioned, though, people prefer units of length over units of frequency (at least I do). Also, the early history of optics focused more on phenomena that were better explained by thinking about wavelength (interference, diffraction gratings, etc.). The twin issues of photon energies and electron levels didn't show up until the early part of the twentieth century. Finally, small as they are, the wavelengths of visible light are measurable, running from 0.4 μm to 0.7 μm. In contrast, the frequencies corresponding to these same wavelengths are on the order of 10^{14} Hz, which are hard to measure. To give you an idea of the magnitude of this frequency, imagine that a second was stretched out to the length of a hundred-year lifetime. A light wave would still go through a complete cycle in 3/100,000 of a (normal) second. So, unless you are working on the border of biology and quantum physics, I would strongly suggest you stick to wavelengths, as I will in this book.

However, one critical issue must be discussed before we put frequency away for good. It involves the fact that light spectra are histograms. Suppose you measure the spectrum of daylight, and that the value at 500 nm is 15 photons/cm²/s/nm. This doesn't mean that there are 15 photons/cm²/s with a wavelength of exactly 500 nm. Instead, it means that, over a 1-nm-wide interval centered on a wavelength of 500 nm, you have 15 photons/cm²/s. The bins in a spectrum don't have to be 1 nm wide, but they all must have the same width.

Let's suppose all the bins are 1 nm wide and centered on whole numbers (i.e., one at 400 nm, one at 401 nm, etc.). What happens if we convert these wavelength values to their frequency counterparts? Let's pick the wavelengths of two neighboring bins and call them λ_1 and λ_2. The corresponding frequencies v_1 and v_2 are equal to c/λ_1 and c/λ_2, where c is the speed of light. We know that $\lambda_1 - \lambda_2$ equals 1 nm, but what does $v_1 - v_2$ equal?

$$v_1 - v_2 = \frac{c}{\lambda_1} - \frac{c}{\lambda_2} = \frac{c(\lambda_2 - \lambda_1)}{\lambda_1 \lambda_2} = -\frac{c}{\lambda_1 \lambda_2} \cong -\frac{c}{\lambda_1^2} \qquad 2.1$$

You can do the last step because λ_1 is close to λ_2. So the width of the frequency bins depends on the wavelengths they correspond to, which means they won't be equal! In fact, they are quite unequal. Bins at the red end of the spectrum (700 nm) are only about one-third as wide as bins at the blue end (400 nm). This means that a spectrum generated using bins with equal frequency intervals would look different from one with equal wavelength intervals. So which one is correct? Neither and both. The take-home message is that the shape of a spectrum depends on whether you have equal frequency bins or equal wavelength bins.

So why should you care? Do everything using wavelengths and it shouldn't matter, right? Unfortunately, not all functions plotted against wavelength are histograms. Visual sensitivity curves, UV damage action spectra—anything of this sort—are known as "point functions" and don't suffer from this bin effect. This can lead to serious misinterpretation. For example, there is a long-standing belief that human visual sensitivity is opti-

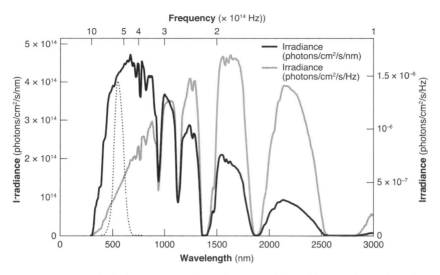

Figure 2.3: The daylight spectrum from figure 2.2 binned by equal wavelength intervals (black) and by equal frequency intervals (gray). The dotted line is the human photopic luminosity curve in response per photon rather than response per Watt.

mized to match daylight. This is based on the fact that our diurnal visual sensitivity curve peaks at 555 nm and daylight peaks at about 510 nm. Sounds like a pretty good fit, right?

First of all, our visual sensitivity curve is always given in units of response per watt, and the daylight spectrum is usually given in watts, despite the fact that our eyes count photons not energy. If you convert our sensitivity curve to response per photon and the daylight spectrum to photons/cm²/s, the peaks are now at 550 nm and about 685 nm, respectively. This isn't as good, but still not a terrible match, right? However, if you measure the daylight spectrum using a light meter that bins by equal frequency intervals, the peak is now at about 1600 nm (Figure 2.3). This is well into the infrared and nowhere near our visual sensitivity curve.

So which is the correct daylight spectrum to compare to our visual sensitivity curve? As we mentioned before, the probability that a photon gets absorbed by a photoreceptor depends on its frequency, not its wavelength, so you could make the argument that the equal frequency bin spectrum is the one to use. However, the correct answer is that comparing the position of peaks is pointless. You can compare the peaks of spectra with one another, so long as they are all taken with equal wavelength bins or equal frequency bins. But you cannot compare them with action spectra and visual sensitivity curves, because both are point functions instead of histograms and so don't depend on whether the x-axis is in frequencies or wavelengths. The only time this is allowable is when all the light has nearly the same wavelength, as is the case in the deep sea. Then the peak stays put (more or less) whether you are working with wavelength or frequency bins. Outside of the deep sea, though, you almost never see spectra this narrow.

So what can you do? It turns out that, while the peak of a histogram depends on how you bin the data, its integral does not. You can prove this with calculus, but instead just think about putting one hundred marbles into ten buckets of varying size. No matter how you split up the marbles between the buckets, you will always have one hundred marbles total. Therefore, if you want to see if human visual sensitivity is optimized for daylight, you look to see if the total amount of light absorbed at all wavelengths is as high as it can be. We will do this in detail in chapter 4, which covers absorption.

An important consequence of the both the photon/energy and wavelength/frequency issues is that you cannot unambiguously say that a light is spectrally neutral (aka "white"). This has nothing to do with the fact that our visual systems perceive a diverse array of spectra as white (called "color constancy"), but instead happens because the way a spectrum looks depends on the units you use and how you bin the data. In other words, a spectrum that has the same amount of energy in each wavelength interval will not have the same amount of photons in each wavelength interval. In addition, both of these spectra will look completely different if you bin by equal frequency intervals. Now, hardly any real light sources come close to being spectrally flat in any of these four arrangements, but it is common in light modeling circles to assume a spectrally flat "white" illuminant. You can of course do this, but remember that it has no universal significance. The spectrum will only be flat for the units and bins that you choose.

IRRADIANCE

So with photons versus watts and wavelength versus frequency out of the way, let's consider what you might actually measure. Unless you work on bio-

Vector irradiance   Radiance   Scalar irradiance

Figure 2.4: The three most common light measurements: radiance, vector irradiance, and scalar irradiance.

luminescence and want to know how much light is being emitted in all directions, you will nearly always be measuring either radiance or irradiance. So, a firm understanding of these two properties goes a long way to clearing things up. To avoid confusion, I will stick to photons, but it is important to realize that radiance and irradiance can just as easily be defined in terms of watts. I will also assume that the measurements are spectral (i.e., when using a spectrometer), but they can just as easily be nonspectral.

While more complicated in principle, irradiance is easier to explain. Simply put, irradiance is the number of photons that hits a small surface over a period of time and has units of photons/s/cm^2 (add a per nm if it is spectral). When you are sitting in front of a fire, the warmth of your skin is roughly proportional to the irradiance striking it. How bright a white piece of paper looks is roughly proportional to the irradiance at its surface.

Irradiance comes in several geometric flavors, depending on the shape of the small surface being hit and how much you care about the angle of the light as it strikes the surface. Except for specialized applications, you only need to consider two types: vector irradiance and scalar irradiance (figure 2.4).

Vector irradiance, by far the most commonly used type, is the number of photons striking a flat surface weighted by the cosine of the angle between the direction of each photon and the perpendicular to the surface. Yes, this sounds awful and arbitrary, but it does make sense. Get a powerful flashlight and point it at your stomach (it helps if you lift your shirt). If the beam is perpendicular to your stomach, it makes a small round spot that feels quite warm. If you tilt the beam, the same amount of light hits your stomach, but the spot is now an ellipse and your stomach doesn't feel as warm (figure 2.5).

Figure 2.5: The same amount of light gets spread over a larger area when it strikes a surface at an angle. This effect, which is proportional to the cosine of the angle of the incident light from the perpendicular, lowers the density of the photons that strike any point on the surface and underlies the cosine rule for calculating vector irradiance.

The area of this ellipse is inversely proportional to the cosine of the angle between the flashlight and the perpendicular to your stomach, so your stomach's warmth is proportional to the cosine of this same angle. So, returning from the analogy, vector irradiance tells you the amount of energy per unit area that the light is imparting to the surface over a given time period. This cosine weighting function also explains why noonday sun feels hotter than early morning sun and partially explains why winter (when the sun is lower in the sky) is colder than summer. Sadly, most light meters don't work like stomachs and special attachments are needed to reproduce this cosine response, which we will get to in chapter 9.

It is critical to remember that vector irradiance depends on the orientation of the surface relative to the illumination. Under the same illumination, the irradiance striking a sideways-facing surface is in general different from that of an upward-facing surface. For no good reason, "irradiance" has generally come to mean down-welling irradiance, which is the vector irradiance striking an upward-facing, horizontal surface (e.g., the ground). This particular measurement has also become identified with the loosely defined term "general illumination." Yes, down-welling irradiance is usually proportional to overall illumination. However, the illumination you care about, such as light entering the eye or bouncing off the side of a fish, may be quite different. For example, the irradiance entering the eye of a person facing west in late afternoon will increase as the sun sets because the sun is moving into a position directly in front of the eye, even though the down-welling irradiance is decreasing. Similarly, using down-welling irradiance to determine how bright the lateral surface of a fish will be is a bad idea. In this case, one should measure horizontal irradiance, which is done by simply turning an irradiance detector sideways.

The second most common type of irradiance, scalar irradiance, does not suffer from the orientation dependence of vector irradiance. Scalar irradiance is given by the number of photons that pass through the surface of a small sphere over a given period of time. It has the same units as vector irradiance, but the directions of the photons do not matter. So, for any location in space there is only one scalar irradiance, while vector irradiance depends on the orientation of the detector.

Despite this great advantage, the only biologists I know of who regularly use scalar irradiance are those interested in the photosynthesis of phytoplankton. Why? I don't know. It could be because scalar irradiance detectors are harder to make (they look like Ping-Pong balls on long black sticks). Also, a detector that collects light from all directions is not useful for determining how much light hits an animal's skin or enters the eye, since the body would normally be shading at least half the field of view. In this case, vector irradiance using the correct orientation is the way to go. However, most irradiance measurements made by biologists are done to determine general illumination levels, which are more accurately described by scalar irradiance. In addition, because scalar irradiance detectors are orientation-independent, you don't have to worry about which way your instrument is facing, which is a real concern when your meter is 500 meters under water, hanging from a cable, and rotating in the current.

The main justification I can see for continuing to use down-welling vector irradiance as a proxy for general illumination is that it has been done for a long time. So, if you want to compare your light measurements to those of previous studies, you need to do the same thing. This is not much different from saying that we should still cut our food with stone knives, but it is something to consider if your work has a historical aspect, such as an examination of long-term changes in light levels in the Arctic. Otherwise I would use vector irradiance when interested in the amount of light striking one side of an opaque object or entering an eye (for example, to determine visual adaptation level), and use scalar irradiance when interested in the brightness of a habitat.

RADIANCE

All forms of irradiance combine light from many directions. Aside from some simple weighting functions, where a photon comes from doesn't matter. However, suppose you need to know the brightness of a particular location—for example the corner of the wing of a butterfly or the dewlap of an anolid lizard. Then you need to use radiance. Like irradiance, radiance is given by how many photons strike a small surface over a period of time, but in this case only photons that arrive from a small set of directions are counted. In addition, the number of photons from that direction is divided by the angular size of the area viewed. This brings us to one of the more intimidating aspects of light measurement, the steradian.

The two-dimensional analog of degree, the steradian (sr), is a unit of solid angle (figure 2.6). I have spent a reasonable portion of my life being confused by steradians (and pretending I wasn't, which didn't help) and finally developed this mental picture. Imagine a circle. From high school geometry, its circumference is $2\pi r$, where r is the radius. The ratio of the circumference of this circle to its radius is then 2π. This is why we say that 360 degrees are equal to 2π radians (and why one radian is about equal to the rather clumsy value of $57.2958°$). Go up a dimension and imagine a sphere. Its surface area is $4\pi r^2$, so the ratio of the surface area to the radius squared is 4π. So we say that a full sphere equals 4π steradians.

Now imagine you are looking at a blue crab swimming by you in the ocean. Depending on its orientation and shape, the crab has a certain cross-sectional area when viewed from your location. The angular area of the crab in steradians is this cross-sectional area divided by the square of the distance between it and you. The radiance of the crab is the amount of light that leaves it and strikes the detector (over a period of time and divided by the size of the detector) divided by the angular area of the crab. Because the full 3D field of view is 4π steradians, one steradian is about 8% of a sphere (roughly the fraction of the globe that Asia occupies). It is also equal to about 3283 square degrees, but solid angles are seldom measured in this way.

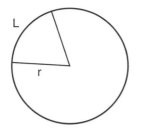 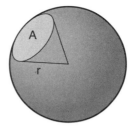

Figure 2.6: Radian (left) and steradian (right). The angle in radians is given by L/r. The solid angle in steradians is given by A/r².

Figure 2.7: Field of view of a radiance detector that is restricted by a tube that is 20 cm long and 4 cm in diameter.

Objects are seldom all one color and brightness, so most of the time you will measure the radiance of a location within an object, not of the whole object itself. In this case, the angular area that you are dividing by is determined by the field of view of your detector. This field of view can usually be figured out via some trigonometry. For example, a small detector at one end of a tube that has a diameter of 4 cm and a length of 20 cm has a field of view of roughly $\pi 2^2/20^2$ or about 0.03 sr (figure 2.7). Underlying all this is the unspoken assumption that the field of view is roughly circular, or at least square. If you have a detector that looks at a long curvy shape in space, I'm not sure what you are measuring, but it isn't radiance.

THE CONSERVATION OF RADIANCE

Perhaps the most often quoted rule from optics is that the radiance of an object decreases with the inverse of the square of its distance. People who have obviously noticed that their friends don't get darker as they walk away nevertheless still insert the inverse square law into their papers. This is a classic example of faith in math trumping basic experience.

The inverse square law is almost always wrong. The following explanation, based on a chapter from Bohren's *Clouds in a Glass of Beer*, explains why. Suppose you are standing five meters away from a small fire on a cold night. Your skin, a pretty good irradiance detector, feels the combined warmth of the fire and the surrounding air. Your eye, an excellent radiance detector, measures the radiance of the fire relative to the dark background. You then move back five meters to a distance of ten meters from the fire. What happens to the irradiance and radiance? Let's assume that the fire emits photons evenly in all directions. The angular area of your body, as viewed by the fire is only one-quarter of what it was before you moved. This is because you are twice as far away, so you appear half as wide and half as tall to a viewer sitting at the fire. Therefore, your body is intercepting only one-quarter of the photons that it had before, so the irradiance is one-quarter of what it was. In personal terms, the fire feels one-quarter as hot on your skin. Spend enough time around campfires or wood stoves and you develop a powerful faith in the inverse square law for irradiance.

But what about the radiance? Just like your body, the angular area of your pupil (as viewed by the fire) is only one-quarter of what it was before and so intercepts only one-quarter of the photons. However, the fire also appears smaller, since you are farther away from it. In fact, the fire has one-quarter of its previous angular area, since it is also half as tall and half as wide. Since you divide the total number of photons by angular area to get radiance, it doesn't change. This simple geometric principle is at the heart of what is called the "conservation of radiance."

There are, however, three caveats. First, the intervening medium has to be "lossless," which means that it doesn't absorb or scatter light. This is true for fairly large distances in air, though not for looking through fog or at distant mountains. This means that the radiance of a lightbulb is about the same whether it is across the street or in your hand. Since outer space has little effect on the propagation of light, it also means that the sun viewed from Pluto has the same radiance as the sun we see. It is much smaller, but just as bright. Water is not a lossless medium, so radiance is only conserved over short distances and only in the clearest habitats.

The second caveat is related to a question that may have already come to mind—if radiance is conserved, then why are the more distant stars dimmer? In some cases, this is due to intervening dust that absorbs light, but much more important is the fact that stars are essentially point sources. Even our closest bright star, Alpha Centauri, has an angular size of only 2 millionths of a degree, about $1/4000^{th}$ the angular size of the smallest object that we can resolve. So, we don't actually see this or any star, instead we see their diffraction patterns. We will talk more about this later, but the important point for now is that, as a star gets farther away, the number of photons from it that enter our pupil decrease, but the image of the star itself does not get smaller. Therefore, there is no balance between the number of photons received and the angular size of the star, so the radiance decreases with the inverse square of the distance. Thus, the second caveat is: radiance is only conserved if the object viewed has not shrunk down to a point.

But how do we decide when an object is a point? Obviously even stars are not infinitesimal geometric points. While it looks like we might have to make an arbitrary decision, there is actually a clear-cut division between extended objects, the radiance of which is conserved, and point objects, where the radiance drops with the inverse square of distance. It all depends on the detector. If the angular size of the object is below the resolution limit of the detector, then it is a point source; if it is above this limit, then it is an extended object.

In a way, this is a shame, because it means that a measurement of radiance, unlike irradiance, depends on who or what is measuring it. Pretty wishy-washy for a physical property. For example, eagles have eyes that are approximately twice as acute as ours. If you had a pet eagle on your shoulder and continued to back away from the small fire we have been talking about, it would become a point source for you before it became one for the eagle. After this point and until it became a point source for the eagle as well, the fire would become dimmer with distance for you but remain at the same brightness for the bird. You would both agree, however, on how much (or how little) the fire was warming you.

The third and last caveat is interesting. If you have ever gone snorkeling or scuba diving, you have noticed (or been told to notice) that things often look bigger under water. This happens because light is refracted at the surface of your mask as it moves from water to air (you can ignore the short trip through the glass of the mask itself). Refraction turns out to be much more interesting than you might suspect, which is discussed further in chapter 6. For now, though, let's just stick with the simple explanation that light bends sharply whenever it moves to a medium with a different refractive index. Due to this bending, an underwater object viewed through a dive mask

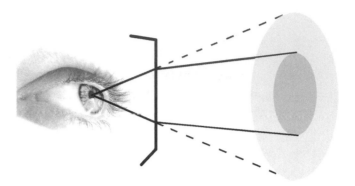

Figure 2.8: Magnification of objects viewed under water due to refraction at the surface of a dive mask. The dotted lines show the apparent size of the circle, the solid lines show the true path of the light, which bends as it passes from the water to the air inside the mask.

looks bigger (figure 2.8). How much bigger the object appears depends on its size and also on how far away it is compared to how far your eyes are from the glass of the mask. Assuming that the mask-eye distance is a typical 3 cm, anything farther away than about 50 cm and less than 15° across in real angular size looks about 33% wider and 33% taller. Think of this as 1.33 times the size and it should ring a bell. Yes, the magnification is equal to the refractive index of water (= 1.33) divided by the refractive index of air (= 1).

Since area equals length squared, the solid angle of objects viewed under water increases by a factor of m^2, where m is the ratio of the indices of the two media ($1.33^2 \cong 1.75$ for water viewed from the air behind your mask). But the total amount of photons that reach your eye from the object has not changed (it can't without violating fundamental conservation laws). Radiance is the total number of photons divided by solid angle, so the radiance of the object decreases by a factor of m^2. In other words, objects viewed under water are larger and dimmer.

Unless you scuba dive like a bloodhound, with your nose to the reef, most things you look at will be at least 50 cm away. Also, most things will be less than 15° across. However, just to be complete, we should think about what happens for closer objects and those larger than 15° across. As an object gets closer to the mask, its magnification decreases until it is exactly one as it

Figure 2.9: Distortion of objects under water due to refraction at an air-water interface (in this case the window of the underwater housing of the camera). Courtesy of Tali Treibitz.

touches the mask. Try this under water the next time you go snorkeling or diving. If you don't dive or snorkel, then stick your hand in a fish tank and move it around. What about large objects? The edges of larger objects (and by "larger" I mean larger in angular size, not real size) get magnified more than m^2-fold, so they start to look stretched and distorted (figure 2.9).

So what happens to the radiance at these close distances and/or larger angular sizes? Does it still go down by a factor of m^2? Amazingly, it still does even though the magnification is not always m^2. It can be proven mathematically or experimentally, but sadly, I can't offer a good intuitive explanation. The underlying reason is that radiance also depends on the angle between the light and the detector, which gets altered as you cross the air-water interface. It turns out that this correction exactly cancels out the variation in magnification and leaves only the m^2 factor.

The end result of all this is that it is not radiance that is conserved but radiance divided by m^2. Going back to our diver, suppose another person sitting on a small boat shines a flashlight beam through the water at him/her. Ignoring the usual units, let's assume that this flashlight emits a beam with a radiance of one in air. As the beam enters the water, its radiance now jumps to 1.75 (= 1.33²). This is what the diver would see if he/she did not have a mask. However, diving without a mask makes you horribly farsighted, so let's put it back on. The mask itself is made of glass, which has an index of about 1.5. So, on the light beam's short trip through the glass of the mask, the radiance jumps to 2.25. Then it hits the air between the mask and the diver's eye, and the radiance drops back to one. So if the diver is wearing a mask, the flashlight looks no brighter or dimmer than it would if he or she were in the boat with the friend. In certain cases though, the radiance actually does change: for example, the diver looking at underwater objects and seeing that they are larger and dimmer. Another case involves diffusely reflecting (i.e., nonshiny) objects in water that are illuminated and viewed from the air above. Bohren and Clothiaux (2006) have a nice photo showing that white plastic spoons in black water-filled bowls look darker. You can see the same effect if you stick your hand in a pond with a dark bottom. I personally think it is fascinating that relatively esoteric geometric discussions sometimes lead to simple observable consequences.

It would be nice to end the chapter on this high note, but I feel that I need to at least address photometric units.

PHOTOMETRIC UNITS

Photometric units are based on human visual perception, derived from decades of experiments in human psychophysics, and are therefore arbitrary when it comes to nonhuman visual systems. Also, as I mentioned in the introduction, photometric units are a nightmare. If you weren't put off by foot-lamberts and nits, try adding apostilbs, skots, and milliblondels (a thousand small blondes?). However, at its core, photometry is based on a fact that is useful to biologists—that vision and other biological processes are not equally efficient at all wavelengths of light. Even within the visible range, human eyes are much more sensitive to 555 nm light than to 410 nm light, which is why yellow looks brighter than violet to us. Therefore, while I would toss the units (unless you plan to work in human visual psychophysics), I would still keep this general principle.

The mathematical implementation of this idea is known as the "weighted integral." One begins with a measured irradiance or radiance spectrum, multiplies the value at each wavelength by a function that corresponds to the

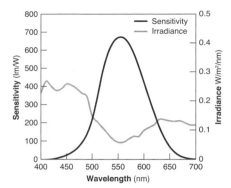 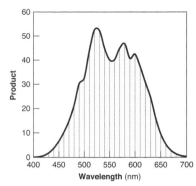

Figure 2.10: Example of a weighted integral from human vision. (Left) Human luminosity curve (black line) and an arbitrary irradiance spectrum (gray line) in energy units. (Right) Product of the two curves. The weighted integral (which in this case is proportional to the perceived brightness of the irradiance in lux) is the area under the curve.

sensitivity of a given eye (or some other biological process), and then integrates. In equation terms, the weighted integral w is:

$$w = \int_{\lambda_1}^{\lambda_2} E(\lambda) V(\lambda) d\lambda,$$
2.2

Where λ_1 and λ_2 define the wavelength range (380 nm to 760 nm for human daylight vision), $E(\lambda)$ is the measured radiance or irradiance in photons or watts, and $V(\lambda)$ is the sensitivity function (either per watt or per photon) (figure 2.10). Remember that the sensitivity function has to be in the same units as the spectrum. In practice, this weighted integral is usually done in Excel or Matlab with discrete wavelength intervals $\Delta\lambda$ of at least 5 nm. In other words:

$$w = \sum_{\lambda_1}^{\lambda_2} E(\lambda) V(\lambda) \Delta\lambda.$$
2.3

$V(\lambda)$ is usually the spectral sensitivity of some photoreceptor determined using microspectrophotometry or the spectral sensitivity of the whole eye using electroretinography. With some knowledge of the visual system of the species you are working with, you can use these formulas to determine how bright an object will appear to that animal. If the animal has photoreceptors of multiple classes, you can determine the relative excitation of each and get a sense of color perception. The same process is often used to assess UV radiation exposure, light available for photosynthesis and other situations where one is dealing with a process that is wavelength dependent.

The big difference between the method I just described and classical photometry is that you start by measuring the entire spectrum and then calculate the integrated terms later. In contrast, photometric sensors usually have filters that mimic the spectral sensitivity of the human light-adapted eye and thus do all the integration within the instrument. It is trivial to add numbers, but it is impossible to recover numbers when only given the sum. Therefore, if you have a light meter that only gives integrated photometric values, sell it and buy a spectrometer.

What if you need to use someone else's data and it is in photometric units? You can convert it to photons if you know the underlying shape of the spectrum. The spectral sensitivity function for human photometric units can be found in many places (and appears in appendix A). Given it and the underlying shape of the spectrum of the measured light, one can work backward from equation 2.3 and determine how many photons there are at each wavelength. There are elegant ways to program this, but I usually just set up a spreadsheet where the measured spectrum is multiplied by some arbitrary value, and then vary that value until the integral weighted by the photometric function matches the measured photometric value.

The original and legitimate use of photometric units is to convert a whole spectrum into one number that is a good estimate for how bright it appears to humans. Suppose you want to show how irradiance changes in ten-minute intervals over the course of a day. You could show hundreds of spectra in some sort of 3D representation, but these take up a lot of space and are difficult to interpret. If all you care about is the general light level, then feel free to give it in photometric units (or in the analog units appropriate for your animal's visual system). The photometric analog of irradiance is illuminance and has SI units called "lux," which are lumens/m². A lumen is one candela multiplied by the angular area of the target in steradians, where candelas trace their origins back to candlepower, which is the light produced by a pure spermaceti candle weighing one sixth of a pound and burning at a rate of 120 grains per hour. You can see why I stay away from these units. At any rate, I do use lux now and then, and a method for converting a spectrum to lux is given in appendix A. Table 2.1 gives you a ballpark feel for the unit. There is also luminance, which is the photometric version of radiance, and

TABLE 2-1
Illuminance under different conditions in lux

Daylight (with direct sunlight)	100,000
Daylight (without direct sunlight)	10,000
Overcast day	1000
Heavy overcast day	100
Late civil twilight	10
Late nautical twilight	1
Full moon	0.1
Quarter moon	0.01
Moonless night (clear sky)	0.001
Moonless night (overcast)	0.0001

trolands, which takes into account the area of your pupil, but they are even more specific to human visual studies.

One last thing about photometric units. Luminance is not the same as brightness. In other words, just because two regions have the same luminance, it doesn't mean they appear equally bright to your eye. Our brain adjusts an image in a number of ways before we actually perceive it. One theory, developed by Dale Purves, is that this is done to minimize the confounding effects of shadows, haze and changing illumination on our perception of the world (Purves and Lotto, 2003). Others have differing hypotheses, but, regardless, the effects are powerful (figure 2.11). In addition, no lesser an au-

Figure 2.11: Squares A and B have the same radiance and luminance but B appears brighter than A. It is thought that this is because our mind recognizes that square B is in the shade and so makes it appear brighter to better match its true reflectance. The vertical lines in the right panel are there to demonstrate that A and B have the same gray value. Courtesy of Edward H. Adelson.

thority than the Optical Society of America states that even isolated mono-chromatic lights of the same luminance do not appear equally bright. For example, a violet light will appear brighter than a yellow light of the same luminance (Smith and Pokorny, 2003). But isn't the luminosity function based on the perceived intensity of monochromatic lights and so shouldn't all monochromatic lights of the same luminance have the same apparent brightness? This bugs me.

In Conclusion

It is a shame that confusing geometry and the proliferation of units has kept so many people away from optics. For biologists not working on humans, it comes down to a few simple rules. (1) Use photons not watts. (2) Use wave-length not frequency, but be extremely careful when comparing the peaks of light spectra to sensitivity curves. (3) Stick to measuring radiance and irradi-ance, but be a pioneer and add scalar irradiance measurements to your pa-pers. (3) Irradiance sometimes obeys the inverse-square law; radiance rarely does. (4) Do not use photometric units, except possibly for lux. Avoid even reading about them. However, the underlying principle of weighted inte-grals is useful for modeling animal vision and other wavelength-dependent processes.

Further Reading

The discussion of radiance and irradiance is based on chapter 15 of *What Light Through Yonder Window Breaks?* by Craig Bohren. The conservation of radiance discussion is based on chapter 4 of Bohren and Clothiaux's *Fundamentals of Atmospheric Radiation*.

Photometric units and general issues of human-based brightness and color perception can be found in *The Science of Color*, edited by Steven Shevell and published by the Optical Society of America.

CHAPTER THREE

Emission

> Yeah we all shine on, like the moon, and the stars, and the sun.
> —*John Lennon* (from "Instant Karma")

Even a casual look at the most boring of surroundings, for example my office, quickly reveals a dazzling diversity of color, sheen, polarization, and other effects. In fact, the diversity of visual phenomena is at least partially responsible for the complex and messy history of optics—for ages people have naturally wanted to catalog and understand what they saw. So it can be surprising to realize that there are few natural sources of light. In fact, there are actually only a few ways of making light, the main two categories being (1) thermal radiation, where light emission is related to the temperature of an object, and (2) luminescence, where light emission is related to specific changes in the energy levels of molecules. Other possibilities, such as light produced when matter meets anti-matter, are not likely to matter in biology. With few exceptions, light in the biological world ultimately comes from two sources, the sun and bioluminescence, which are exemplars of the two main mechanisms of light emission. So, let's being with the sun, which is a classic example of a thermal radiator.

THERMAL RADIATION

Thermal radiation can confuse people, but is actually a straightforward concept with a few simple rules. First, everything radiates. People, snow, even liquid helium at 4°K—they all emit radiation. This radiation is a natural consequence of the thermal motion of the electrons that make up the object. Ultimately, all electromagnetic radiation, thermal or not, comes from the motion of charged particles (technically it comes from the acceleration of charged particles, but in any realistic situation, rapidly moving charged particles are changing directions many times and so are also accelerating).

Second, the amount of radiation emitted by an object goes up quickly with temperature, due to the increased motion of the particles within it. In watts, the total radiation emitted by an object (integrated over all wavelengths) is proportional to the fourth power of its absolute temperature. Measured in photons, it is proportional to the third power of the temperature. So, while liquid helium (T = ~4° K) does emit radiation, it emits far less than we do, about 30 million times less in watts. Before you discount this, consider that the cosmic microwave background radiation, which has been measured often and is the cornerstone of modern cosmology, is even fainter (T = ~2.7°K).

Third, objects also absorb radiation (more about this in chapter 4). The fraction of incident radiation they absorb is known as their "absorptivity." The absorptivity of macroscopic objects cannot be higher than 100%, but what about the relationship between absorption and emission?

If an object absorbs more than it emits, it gets hotter. If it absorbs less than it emits, it gets colder. Sounds simple, and it happens every day. Lie outside on a sunny day and feel your skin warm up. However, what if an object were somehow more efficient at absorbing radiation than it was at emitting it? We'll define "emitting efficiency" in a moment, but for now, let's call this efficiency emissivity and assume we know what it means.

Imagine that you own a purple ball that is a more efficient absorber than emitter. In other words, its absorptivity is greater than its emissivity. Put this ball in a glass of water that has the same temperature. Over time, the ball's temperature will go up and the water's temperature will go down. This breaks the second law of thermodynamics and allows you to do illegal things like cool your refrigerator just by putting your magic ball inside it. Therefore, an object in equilibrium with the environment has to absorb as much radiation as it emits. Another way of saying this is that absorptivity has to equal emissivity.

Of course, absorptivity and emissivity, like everything in optics, depend on wavelength. So, does the absorptivity have to equal the emissivity at every wavelength, or can they actually have quite different wavelength dependencies just so long as the integrated values over the entire wavelength range are the same? For example, can a rabbit have a higher absorptivity than emissivity in the visible range and make up for it by having a higher emissivity than absorptivity in the infrared portion of the spectrum? Surprisingly, it cannot. Something known as the Principle of Detailed Balance requires that absorptivity equal emissivity at each wavelength. The proof involves subtle arguments about the time reversal of physical processes, so we will just assume that it is true (see Reif, 1965, for details).

The equality of absorptivity and emissivity has some interesting consequences. For example, a perfectly reflective object absorbs no radiation—all of it just bounces away. So its absorptivity is zero, which means that its emissivity is also zero. This is why tea kettles and thermoses are often shiny. The high reflectivity means a low emissivity, which reduces radiative (but not conductive) heat loss, though I am sure style has something to do with it as well. It also explains those flimsy metallic space blankets that you find in camping stores. They are lousy insulators, but are good at reducing radiative heat loss to the sky.

Oddly, no terrestrial animals seem to take advantage of this principle. I don't know of any terrestrial animals that are covered with mirrors or otherwise have high infrared reflectance that would lower their potential radiative heat loss. There are many mirrored aquatic animals, but heat transfer under water is dominated by conduction via the water, and infrared radiation barely penetrates, so I doubt their shiny surfaces affect their temperature. Arctic animals are often white, but only in the visible range. In the infrared, where most of the radiative exchange occurs, they are nearly black. Terrestrial plants however, have infrared reflectances close to 100% (figure 3.1). This has usually been thought to reduce solar heating of the leaves, but perhaps it also serves to reduce radiative heat loss at night. This would be interesting to test.

It is easy to define the 100% point for absorptivity. It occurs when all the incident radiation is absorbed. But how do you set the 100% point for emissivity? Since absorptivity equals emissivity, a 100% emissive object absorbs all radiation that hits it and, because of this, is called a "black body radiator." A black body radiator in this formal sense emits the maximum amount of radiation possible for its temperature. It also has a characteristic radiance spectrum $L(\lambda)$ that depends only on its temperature (Planck, 1901):

$$L(\lambda) = \frac{2c}{\lambda^4} \frac{1}{e^{\frac{hc}{\lambda kT}} - 1} \qquad 3.1$$

If this equation impresses you, it should. As Bohren and Clothiaux (2006) note, it was derived by combining thermodynamics with the hypothesis that light comes in discrete units and includes the speed of light c, Planck's con-

Figure 3.1: Tree and grass photographed under infrared light. Note the high reflectance of the leaves. Courtesy of Seng P. Merrill.

stant h, and Boltzmann's constant k—the central constants of relativity, quantum mechanics, and thermodynamics respectively. By the way, if you have seen this equation before, don't reach for your Wite-Out. Planck's equation is usually given in energy units, mine is in photons (recall that the energy of a photon is hc/λ). It is also giving you radiance instead of the more typical radiant excitance (see appendix F for various other forms of this equation).

Before we look at the radiation emitted by black bodies of different temperatures, there are a few things to keep in mind. First, as we said before, absorptivity and emissivity depend on wavelength. If they didn't, the world wouldn't be so colorful. The electromagnetic (EM) spectrum covers a gigantic range, all the way from superenergetic gamma rays to radio signals whose wavelengths can be miles long. So, even objects that look colorless to us may have wildly different absorptivities at nonvisual wavelengths. Water is a good and important example. While fairly transparent and colorless in the visible range, at shorter and longer wavelengths it rapidly becomes opaque (figure 3.2). Since most organisms live in water and all organisms are full of water, this narrow window of low absorptivity is the only thing between us and blindness. It also means that, over most of the EM spectrum, organisms are

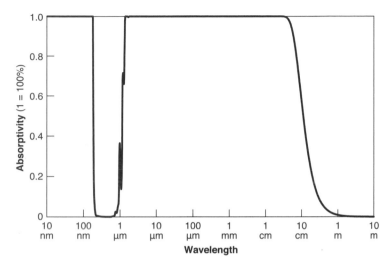

Figure 3.2: Fraction of light absorbed after passing through 1 cm of water. Note the narrow window of low absorption in the visible range and a second window at wavelengths above 1 m. Based on data compiled by Segelstein, 1981.

black bodies, absorbing almost all the radiation that hits them. However, although it is commonly said that all people look black under infrared film, most IR photography techniques are only sensitive to near infrared (700–900 nm). At these wavelengths, pale-skinned people still look pale. You need to image at wavelengths above 1.6 μm for everyone to look black (figure 3.3).

The second important thing to keep in mind is that absorptivity doesn't just depend on the material; it depends on structure and the resulting internal path length of the light. In other words, a black body doesn't have to be black. In fact, the archetypal black body radiator is a spherical cavity carved into relatively shiny metal. Light enters the cavity through a small hole and then bounces off the curved walls so many times that it eventually gets absorbed, even if the absorptivity of the metal itself is low. As physicists like to say, there is nothing blacker than a hole. Similarly, animal tissue, due to its internal complexity, sends photons on a merry bouncing ride that vastly increases their chance for absorption. The open ocean, due to its great depth and scattering properties, lets out very few of the photons that make it through the surface, essentially swallowing the EM spectrum whole. Even in the visible range, less than 3% of the photons that enter make it back out. This is all discussed in more detail in chapters 4 and 5, but the central point

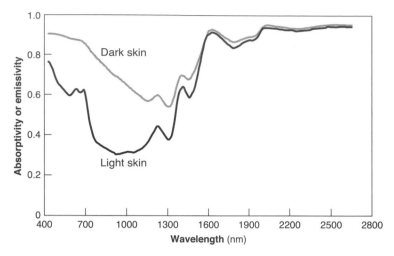

Figure 3.3: Absorptivity and emissivity of human skin at visible (400–700 nm) and near infrared wavelengths. Based on data from Jacquez et al., 1955.

is that, while a black body radiator is ultimately a platonic ideal, most of the biological world approximates it well. The water content and complexity of biological structures ensures that the only portions of the EM spectrum where they might not act like black bodies are from the near-UV to the shortwave-IR (350–2.5 um), with plant leaves being the main exception for wavelengths longer than 1.5 um.

With these two issues out of the way, let's look at the implications of Planck's equation (3.1). Though it is not immediately obvious, as a black body gets hotter, its wavelength of peak emission gets shorter (figure 3.4). In loose terms, heat something, and the light it emits will get "bluer." The wavelength of peak photon emission for a mammal (T = ~310° K) is about 12 µm. Remember though, that the location of the peak of any spectrum depends on whether you bin your spectra by equal wavelength intervals or equal frequency intervals. If you bin by frequency, the peak wavelength is 30 µm. We'll stick with wavelengths throughout this book, but this is another example of why obsessing about the location of spectral peaks is unwise, unless you are comparing spectra with other spectra (and not with visual sensitivity curves or photochemical action spectra).

Also, it is critical to remember that heating a black body increases its emission at all wavelengths. For example, take a red-hot horseshoe and heat it some more. It will turn orange, but that doesn't mean its emission in the

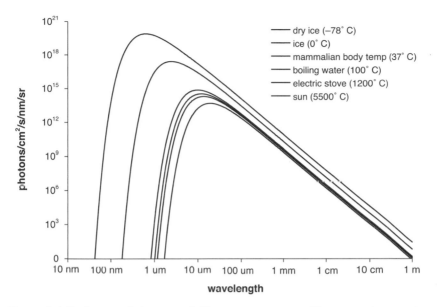

Figure 3.4: Radiance of objects at different temperatures. The curves increase in height with increasing temperature.

red has gone down. In fact, its emission in the red has gone up by a lot. Seems obvious, but you often hear people saying that mammals emit in the infrared and that the sun emits in the visible. The sun also emits in the infrared—far more than mammals do. If you don't believe this, try heating the earth with a sun-sized pile of gerbils. (As an aside, though, if you account for the surface:volume effects of heat dissipation, a sun-sized pile of gerbils would be far hotter than the sun, because the power output of the sun is only about 27 watts/m^3, far less than the roughly 800 W/m^3 of an adult human male. As my dad says, the sun is a slow cooker. The vast difference in temperature is a surface:volume effect writ large. For its volume, the sun has far less surface area through which to release heat, so it gets very hot.)

Keeping all these caveats in mind, the wavelength of the peak photon emission in μm for any black body radiator is 3680/T. The wavelength of peak energy emission is slightly different, equal to 2898/T. In both cases, the temperature must be in Kelvins.

Going back to the radiation from our mammalian body, either 12 μm (using equal wavelength bins) or 30 μm (using equal frequency bins) is in the middle of the infrared band. The radiance at this 12 μm peak is about

5.4×10^{13} photons/s/cm²/nm, which is a lot of photons, about equal to what a 1200°C electric stove puts out in the red region of the spectrum. The spectral distribution for black body radiation of any temperature is quite broad, though. In the case of mammals, the radiance is over half its peak value from 7 μm to 26 μm. Mammals also put out a little bit of visible radiation (too dim for us to see) and a long tail of microwaves and radio waves. Mammals even emit an extremely small amount of ultraviolet radiation, approximately one photon/cm²/sr about every 10^{30} years. If you include all wavelengths, a large adult male with a surface area of 2 m² and a typical skin temperature of 33°C emits about 1000 watts, almost the same as a decent hair dryer and ten to twenty times our basal metabolic rate. If it weren't for the fact that we are also absorbing nearly as much radiation from our surroundings, we would quickly freeze to death.

Before we get to the sun itself, one last confusing term concerning thermal radiation must be mentioned—color temperature. Pretend that you have just bought an amazing electric stove that you can heat to 30,000°K. You take it home and turn it on. Once the temperature gets to about 1000°K, you will notice that the burner is glowing a deep red (if your eyes were dark adapted you would be able to see this glow at about 650°K). At 3000°K, it will be a slightly orange-yellow. At 6000°K, it will look white and be unbearably bright (as bright as the sun). Above this temperature, it will remain white, but you will notice an increasingly blue tint. At 30,000°K, the burner will still be white, with a touch of violet, and your house will be burning down.

This correlation between temperature and color, and the heavy use of natural and artificial incandescent sources in photography, inspired the development of the concept of color temperature by lighting engineers. The color temperature of a light is simply the temperature of a black body radiator at which the black body spectrum best matches the light's spectrum. For incandescent lights or the sun (which are both roughly black body radiators), the color temperature closely matches the actual physical temperature of the light source. For example, the color temperature of a tungsten bulb when the filament is heated to 3000°K is close to 3000°K. For non-incandescent sources like fluorescent bulbs and blue sky, it is a bit trickier, but still boils down to finding the best match. In these cases, it is important to remember that the color temperature says nothing about the actual temperature of the object. For example, the color temperature of the blue sky is usually over 12,000°K, but its radiative temperature is only about 250°K (= −23° C).

If you do any photography in your research, color temperature is worth understanding, because it varies dramatically. The color temperature of natural illumination runs from about 4000°K (moonlight) to essentially infinite for deep twilight. Artificial illumination covers an even greater range. Our eyes, and those of many other species, adapt to changes in the color temperature. This is called "color constancy" and allows us to see a piece of white paper as white under a tremendous range of irradiances and color temperatures. Digital cameras try to adapt via what is known as "white balancing," but often do a poor job. Film cameras don't adapt at all, which is why there are different kinds of film for different lighting situations. For this reason, if you want to compare a set of color photographs with one another, it is critical that you find some way to compensate for this. By far the simplest way to do this is to put a small swatch amidst a set of colors in your images, just as geologists always put a coin or a hammer in pictures of their rocks, for scale. More about this in chapter 9.

SUNLIGHT AND SKYLIGHT

I was fascinated with mythology as a child, and couldn't help noticing that nearly every culture put their sun god at or near the top of the heap. The Greeks were an exception. Having spent many days in high school being flattened by the Mediterranean sun, I still can't understand why the Greeks gave preeminence to thunderstorms. Even the most atheistic biologist can't help but be awed by a disk only ½ a degree across that can't be looked at, but is ultimately the power source for the entire biosphere. Even deep-sea vent fauna require oxygen for their chemosynthetic metabolism—oxygen ultimately produced by light-harvesting phytoplankton.

The sun is a classic example of a black body radiator. Due to its low reflectivity and enormous size, it absorbs nearly all the radiation that strikes it and radiates with a color temperature of about 5800°K. About one-third of the total radiation from the UVB to the shortwave infrared (280 nm–4 μm) is absorbed or scattered into space by our atmosphere, with the big cuts occurring in the UV (due to ozone) and in portions of the infrared (due to water vapor, CO_2 and O_2) (figure 3.5). However, big chunks of the infrared spectrum get through the atmosphere nearly unscathed, a fact that has not been overlooked by astronomers and the makers of earth-observing satellites.

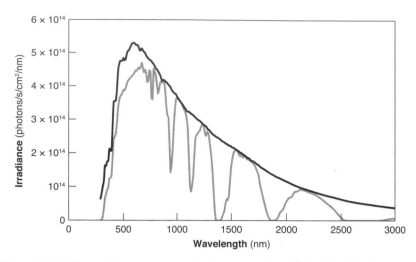

Figure 3.5: Solar irradiance just outside the atmosphere (black) and at the earth's surface (gray). Based on the standard reference spectra measured by the American Society for Testing and Materials.

So, as far as we are concerned, the sun is a 5800K black body source about half a degree across, whose illumination is cut in the UV and parts of the IR due to absorption and scattering by gases in the atmosphere. This is all we need to know about it. If one morning, the sun shrank to the size of the moon and moved to a lunar orbit, 99.9999% of the life on this planet wouldn't know the difference. The only obvious difference would be its new path in the sky and some weird changes in the tides.

It is not often appreciated how small the sun is. One-half a degree is about the angular size of your thumbnail if your arm is fully extended. It is also only about 1/100,000 of the angular area of the sky. Therefore, while the radiance of the sky is far less than that of the sun, its relatively enormous size means that the sky's contribution to down-welling irradiance is substantial at certain wavelengths. As you might guess from the color of the sky, its contribution is greatest at the lower wavelengths. In the blue region of the spectrum (400–500 nm), the sky contributes about 15%–25% to the total down-welling irradiance. In the UVA (320–400 nm), it contributes 25%–50%, and in the UVB it contributes 50% to nearly 100% (figure 3.6).

The high contribution of skylight to the irradiance in the ultraviolet partially explains why people get so sunburned in open places like beaches, fields, and lakes. In all these habitats, there is substantially more visible sky

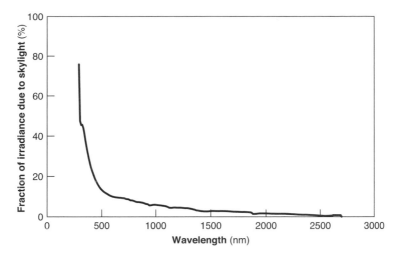

Figure 3.6: Fraction of solar irradiance (at earth's surface near noon) due to sky-light alone. Based on the standard reference spectra measured by the American Society for Testing and Materials.

than is found in typical urban and suburban landscapes. The importance of the skylight to total UV also suggests that sitting under a small umbrella at the beach may not save you from sunburn, even if you are out of the direct sun. You will feel cooler, because the skylight contributes almost nothing to the infrared irradiance, but your dose of UVB photons (those responsible for sunburn) may only be cut by half.

The contribution of skylight to the total irradiance at lower wavelengths increases as the sun goes down. This is partially because the sun's radiance drops as it approaches the horizon, due to increased absorption and scattering that occurs over its longer path through the lower atmosphere. You can barely glance at the noonday sun, but can often easily watch the setting sun. Also, as anyone who watches sunsets knows, its radiance is quite small at the lower wavelengths (i.e., the setting sun is red), so it contributes little to the down-welling irradiance in the blue and UV. Even if the sun's radiance were not affected by our atmosphere, its contribution to the total irradiance would drop as it set due to the cosine law of irradiance. For all these reasons, and several others, down-welling irradiance shifts to shorter wavelengths when the sun is at or just below the horizon (figure 3.7). Because of this, biologists have been interested for quite some time in how color vision and associated behavior change in response to these dramatic changes in irradiance.

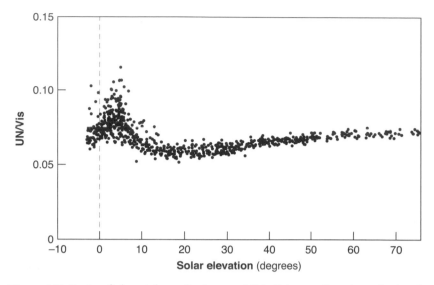

Figure 3.7: Ratio of ultraviolet radiation to visible light as a function of solar elevation on clear days in Granada, Spain. Dotted line denotes sunset. Data courtesy of Javier Hernandez-Andres and Raymond Lee.

The setting and rising of the sun brings us to an important fact. Despite Copernicus, as far as life on earth is concerned, the sun moves. Its apparent position goes through 360° in twenty-four hours, so it moves one degree every four minutes. How much of that motion is horizontal and how much vertical depends on your latitude and the time of year. A good rule of thumb is that, near sunrise and sunset, the sun drops (or rises) *roughly* one degree every $4/(\cos\theta \sin\phi)$ minutes, where θ is the current elevation of the sun (0° is sunset) and ϕ is the highest elevation the sun gets on that day (a more exact formula for all times of day is given in appendix F). Due to the 23.5° tilt of the earth (figure 3.8), ϕ depends on season. It is highest during the summer solstice (= 90° + 23.5° latitude) and lowest during the winter solstice (= 90° −23.5° latitude). Averaged over the whole year, ϕ equals 90° minus your latitude.

So, near sunrise and sunset in my home town of Chapel Hill, North Carolina (latitude = 36°), the sun drops or rises a degree about every five minutes on average (figure 3.9). In the winter it takes eight minutes to drop or rise by a degree, and in the summer four. If I lived on the equator, the sun would drop a degree every 4–4.36 minutes, depending on season. This is

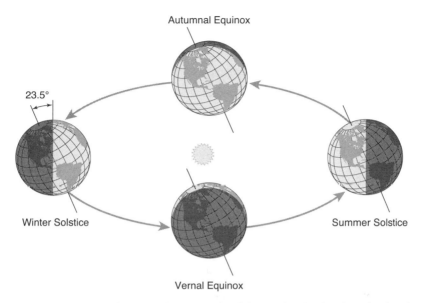

Figure 3.8: Diagram showing the 23.5° tilt of the earth, which affects both solar and lunar elevation as a function of season.

why northerners on vacations in the tropics often talk about night falling like a wall. If I lived in Anchorage, Alaska (latitude = 61°), the sun would take five minutes to drop a degree in the summer and a whopping forty-two minutes to drop a degree on the winter solstice. Winters in the polar regions are all about twilight.

Regardless of where you live, down-welling irradiance is surprisingly constant over most of the day (figure 3.10). Once the sun is more than 10° above the horizon (about fifty minutes after sunrise or before sunset in Chapel Hill), the irradiance on a clear day varies by at most tenfold. Generally it varies less, because, unless you live in the tropics, the sun is never actually directly overhead. In Chapel Hill, the peak height of the sun ranges from 30°–77°, with an average of 54°. So, even in the southern United States, the irradiance only varies on average by a factor of five over most of the day. At higher latitudes in the winter, the variation is even less, especially when you add the typically overcast skies you'll see. Most research on light entrainment of circadian rhythms has focused on the sharp changes in illumination that occur at dawn and dusk. However it would be interesting to know whether the relatively constant illumination levels during winter at polar latitudes af-

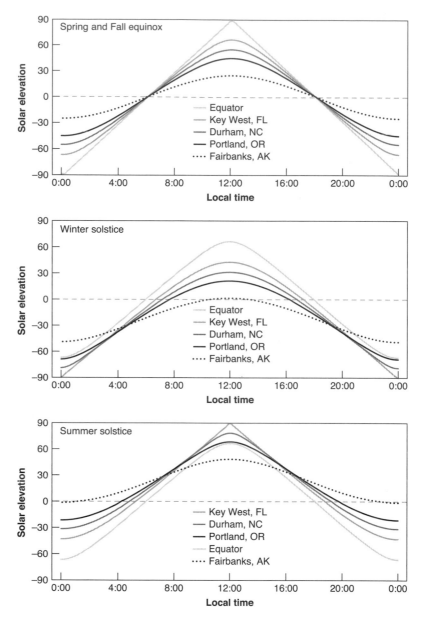

Figure 3.9: Solar elevation during the winter solstice and spring and fall equinoxes for four locations at latitudes 0°, 23.5° (Tropic of Cancer), 35°, 45°, and 65° N.

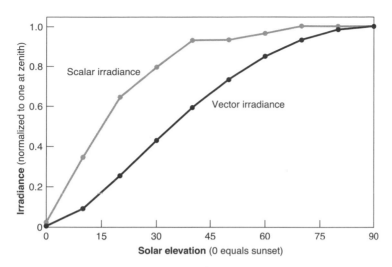

Figure 3.10: Vector and scalar irradiance (at a wavelength of 560 nm) on a clear day as a function of solar elevation.

fect circadian rhythms. The tremendous number of coffee and wine shops I see whenever I visit my colleagues in Lund, Sweden (latitude − 56°), have often made me wonder if caffeine and alcohol are being used as a substitute for the variation in solar irradiance that our species appears to prefer.

TWILIGHT

Once the sun drops below 10°, irradiance decreases rapidly. In Chapel Hill, it drops by a factor of about twenty between 10° and sunset, and by about a millionfold over the next hour as the sun drops a further 15° (figure 3.11). Due to this extreme drop, twilight has been divided into three parts. During civil twilight, the sun is between 0° and 6° below the horizon. If the sky is clear, it is a deep blue. If there are clouds, depending on their location and height, they will be lit by the sun (which is not yet below the horizon when viewed from that height). True to its name, there is enough light to perform most daylight activities (Leibowitz and Owens, 1991), and, as such, its limits are connected with laws about flying and headlight use. Looking upward, one can usually see the brighter planets and the brightest stars.

During nautical twilight, the sun is between 6° and 12° below the horizon and irradiance drops rapidly, about three- to fourfold for every degree.

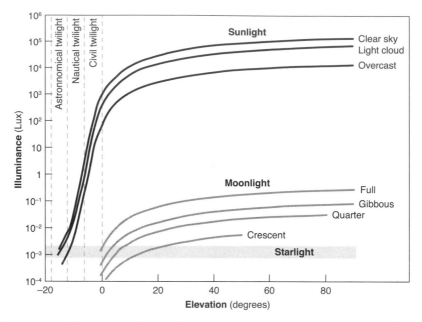

Figure 3.11: Illuminance due to the moon, sun, and stars at various times of day. Based on data from Bond and Henderson, 1963.

At the beginning of nautical twilight, the sky is still blue. By its end, most people see the sky as black and consider that night has fallen. The term "nautical" comes from the fact that, at sea, you can just make out the horizon when the sun is 12° below the horizon.

The final third of twilight, astronomical twilight, is indistinguishable from night by most humans. During this period, the sun is between 12° and 18° below the horizon. While irradiance still decreases, it does so at a slower pace. During this period, the fainter astronomical objects such as nebulae and galaxies become visible.

In addition to the extremely rapid decrease in irradiance, nautical and late civil twilight are also distinguished by their extraordinary blueness. This may not seem surprising. After all, the sun is gone and so you would expect the irradiance to be dominated by the blue skylight. However, it turns out that the late twilight sky is blue for a different reason than the daytime sky is. The blue of the daytime sky is due to scattered sunlight, something that has been pounded into most of us by age ten. There are subtler aspects to this that will be discussed in chapter 5, but the basic premise is true. However, once the sun drops significantly below the horizon, its direct light can only reach the upper atmosphere. At the end of civil twilight, for example, the sky directly overhead (the zenith) only receives direct sunlight at altitudes of 40 km or greater (Bohren and Clothiaux, 2006). At this altitude, the density of air is less than 1/1000th of what it is at sea level and therefore scatters little light. The direct sunlight that does make it to the zenith sky during twilight has also passed through a lot of atmosphere. Therefore a lot of the shorter wavelengths have been scattered away (this is the same process that makes the setting sun red). It turns out that if scattering were the whole story, the twilight sky wouldn't be blue, but a pale yellow (Hulbert, 1953).

Several years ago a colleague of mine, Almut Kelber, discovered that certain species can see color at night (Kelber et al., 2002). The prize goes to the elephant hawkmoth, *Deilephila elpenor*, which can discriminate colors under dim starlight—at light levels in which we can barely see at all. I became curious about what the twilight and nocturnal world looks like to animals that can still see color and started compiling and measuring spectra. Unfortunately, while spectroscopic equipment has improved greatly over the years, light pollution has increased as well, which made everything harder. The twilight spectra were taken by my graduate student Alison Sweeney from a remote beach on a barrier island in North Carolina. The twilight was followed by an intense wind storm, and my student, not the largest of people, had to hug the sand from within her tent to keep the tent from rolling down the beach like a party balloon. She came back frazzled, but with data in hand. Alison, my first student and now a colleague, appears a few more times in this book. One of the best parts of being a teacher is watching your students grow up.

As you might suspect, she found that the twilight sky was blue. However, the spectrum was quite different from that of a daylight sky. Instead of a steady increase in irradiance with decreasing wavelength (like you see in a daylight blue sky), the spectrum looked like a large chunk had been taken out of it (figure 3.12). It turns out that this chunk, centered at about 600 nm, is due to absorption of visible light by ozone.

What we saw had been seen before, and was best described by Hulbert (1953). Ozone, mostly found between 10 km and 50 km above sea level, is famous for absorbing UV radiation, particularly the nasty stuff below 300 nm. The Antarctic ozone hole and the chlorofluorocarbons that at least partially caused it are household words. Without the ozone layer, which would

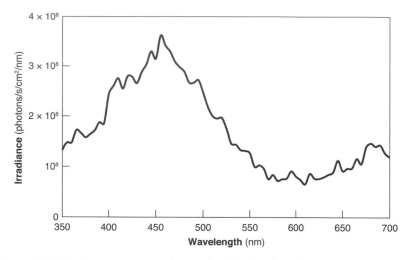

Figure 3.12: Irradiance on a North Carolina beach when the sun was 10.6° below the horizon (late nautical twilight). Data from Johnsen et al., 2006.

only be a few mm thick at sea level pressures, it is doubtful that life would ever have left the water. However, ozone also has a wide absorption band in the visible range, known as the "Chappuis band" (after James Chappuis, who made liquid ozone in 1882 and found that it was deep purple[Taylor, 2005]). Ozone's absorption of visible light is a thousand times weaker than its absorption of UV radiation.

So, during the day it has almost no effect on the color of the sky. When the sun is below the horizon though, its light takes a long raking pass across the ozone layer (figure 3.13). By the time the light reaches the overhead sky and is scattered back down to us, at least half of the orange, yellow, and red light has been removed, leaving blue and violet. So, the similarity between the daytime sky and the twilight sky is a coincidence. If the Chappuis band were shifted to shorter wavelengths, our twilights would be a deep red. Another way to distinguish the blue of late twilight from that of daytime is to look at an overcast twilight sky. Overcast daylight is gray, but overcast twilight is still a deep blue. This is because the absorption of the longer wavelengths occurs at altitudes far above the clouds. A nice semipopular discussion of this and all the subtleties of the color of the sky can be found in Götz Hoeppe's *Why the Sky Is Blue*.

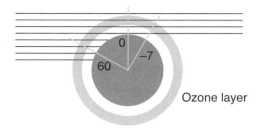

Figure 3.13: Passage of sunlight through the atmosphere and ozone layer for different solar elevations. When the solar elevation is 60°, the path through the ozone layer is short. However, when the solar elevation is −7°, the path through the ozone layer is long, greatly attenuating the long-wavelength light, making late twilight blue. After Bohren C. F. and E. E. Clothiaux, 2006.

MOONLIGHT

The lit side of the moon has a temperature of about 400K, so it does emit some radiation (about as much as a boiling pot of water). However, the vast majority of lunar radiation at visible wavelengths is reflected sunlight. Because of this, a sky lit by the full moon looks much like a sky lit by the sun, though 500,000 times dimmer. Digital cameras are now so sensitive that you can easily prove this to yourself. A two-second exposure using a $100 f1.4 lens with your camera speed set to ISO 1600 will do it. The moon will look white (and overexposed), the sky will be blue, and the landscape will look like you shot the image during the day.

However, everything will look a little bit yellower because the moon reflects about twice as much red light (700 nm) as blue light (400 nm) (Lynch and Livingstone, 1995). The fact that we don't see the slight yellow-brown color of the moon has always bothered me. It is true that objects lit by moonlight are usually too dim to excite our cones, but the moon itself is much brighter. We can certainly see the reddish color of the moon during a lunar eclipse. I have also seen photographs of the moon that show it to be brown.

The apparent path of the moon does not exactly overlap that of the sun. Otherwise we would have a lunar eclipse every month. However, the paths are similar enough for most biological purposes. Also, except during twilight (when the sky can be lit by both moonlight and scattered sunlight), the relative brightness of the moon and the sky is just like that of the sun and the sky.

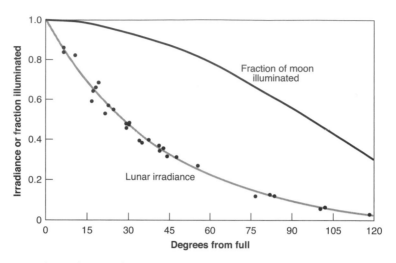

Figure 3.14: Irradiance of moonlight and fraction of moon illuminated as a function of degrees from full moon (0° is full moon; 180° is new moon). Data from Lawrence et al., 2003.

So you can use everything you know about light levels during sunrise and sunset for moonrise and moonset. In other words, the moonlit sky is just a sunlit sky with an economy lightbulb.

The one complicating factor is lunar phase. While sunspots and such have their effects, the irradiance of the sun at a given elevation is relatively constant. The irradiance due to moonlight, however, varies with phase. You might at first suspect that it varies in proportion to the fraction of the moon that is lit (as seen by us). However, the variation is much larger. For example, the down-welling irradiance during a half moon is not half the irradiance under a full moon, but only 1/10 of it (figure 3.14). This means that the average radiance of the lit portion of the moon is not constant. In other words, when less of the moon is lit, the part that is lit is also dimmer.

In fact, the average radiance of the lit part of a half moon (phase angle of 90°) is only about 1/5 that of a full moon. If you become an astronaut, you would be even more impressed, because then you could see a truly full moon. You cannot see a perfectly full moon on earth because the earth gets in the way and you have a lunar eclipse. So the fullest moons we ever see are actually about five degrees from truly full. Astronauts who have placed themselves directly between the sun and moon have found that the moon increases in radiance by another 40% over these last few degrees. In short, the moon is like a retroreflector on a bicycle, sending a great deal of light directly back to where it came from.

This retroreflection, known as the "lunar opposition effect," is primarily due to two things. First, the moon is lumpy. It has mountains and is pocked with craters. As the moon becomes less full, these craters and mountains get lit more and more from the side, which means that they cast longer shadows. This is important to amateur astronomers because a side-lit half moon looks much more dramatic under a telescope than does a full moon, just like mountain landscapes on earth look more dramatic during sunrise and sunset. This also happens on a smaller scale with the particles of the lunar soil itself. Second, it appears that the lunar soil may act a bit like a sheet of glass and have a shiny reflection on top of its main diffuse reflection (Helfenstein et al., 1997). This shiny reflection, known as "coherent scattering," will be discussed more in chapter 6. The bottom line for now, though, is that the phase of the moon has a big effect on lunar irradiance.

What makes this all even more dramatic is that full moons get much higher in the sky than do half and crescent moons—though this is not quite true. Half and crescent moons also get high in the sky, but during the day. The less full a moon is, the closer it is to the sun. Because it is so close to the sun, a crescent moon is highest around noon, when the sun is also highest. Half moons are 90° from the sun and so are highest at dawn and dusk. During the night, both crescent and half moons are often either below the horizon or so close to it that they are obscured by the landscape. Full moons rise when the sun sets, set when the sun rises, and are highest in the middle of the night. Because they are directly opposite the position of the sun, their peak height matches that of the sun in the opposite season. Therefore, the full moon is highest in the winter and lowest in the summer.

The upshot of all this (and why I am burying you in astronomy) is that the variation in nocturnal illumination is much greater than you would expect from how much of the moon is lit, especially if you are considering illumination late at night or the total illumination integrated over a whole night. If you ignore starlight, the total dose (irradiance × time) of moonlight over the course of a winter night is about 5000 times greater for a full moon than for a crescent moon. If you include the background starlight, the difference is still in the range of 50–200-fold.

All this used to be second nature to people, but these days I think the phases and positions of the moon are mostly noticed by late-night dog walkers. Many animals however, do take notice. As is well known, many species, particularly aquatic ones, spawn during certain phases of the moon (reviewed by Korringa, 1947, and Giese, 1959). The timing relative to phase can be quite varied. Some spawn during the new moon, some during the full moon, some, such as many corals, spawn a few days after the full moon. Others spawn on particular full or new moons, combining seasonal timing with lunar phase.

There is a massive literature on lunar phase and biological rhythms that I won't duplicate here. However, it is important for people to realize that there is more information in lunar irradiance than might be expected. For example, outside the tropics, and barring clouds, the brightest night of the year is reliably the full moon nearest the winter solstice. So whether you have a dog or not, go for a walk on that night in late December or early January. You will be impressed.

MOONLESS NIGHTS

As we just mentioned, the moon can be above the horizon all night or for none of it. On average though, it is below the horizon for about half the night. Although the light that remains during moonless nights is quite dim (about one to two orders of magnitude dimmer than moonlight), it is just bright enough to see by. It is also a fascinating concoction of emitted and reflected light from both our atmosphere and outer space. The three big contributors are integrated starlight, zodiacal light, and airglow. Integrated starlight, as you might imagine, is the light from all the stars. However, most of this light is from stars you cannot see, even from the most pristine rural sky. The average integrated starlight is about equal to that of one thousand first-magnitude stars (Matilla, 1980). First-magnitude stars, like Antares and Betelgeuse, are fairly rare. You can usually only see about eight stars this bright at any given time. True, we also see lots of dimmer stars as well, but even if you took all these away, the irradiance from integrated starlight would not drop by much. The real contribution comes from small red stars that are at least a hundred times dimmer than what you can see with the naked eye. These billions of invisible stars, mostly in or near the Milky Way band, make the nighttime sky redder than you might imagine. Because most of the stars

Figure 3.15. Zodiacal light. Courtesy of Y. Beletsky, European Southern
Observatory.

are in or near the Milky Way, the contribution of integrated starlight to
down-welling irradiance depends on how high the Milky Way is in the sky.

A second major contributor to down-welling irradiance on moonless
nights is zodiacal light. This is sunlight scattered by dust in the solar system.
Its name comes from the fact that the constellations of the zodiac demarcate
the plane of the solar system in our sky. Since it is scattered sunlight, zodiacal
light is fairly white. It is generally brightest on the horizon directly over
where the sun sets (or is about to rise), but can brighten a large portion of the
sky (figure 3.15).

The final significant light source is airglow, which is entirely terrestrial
(Lynch and Livingston, 1995). While it is a complicated phenomenon that
occurs during both day and night, the primary source at night is light from
the recombination of oxygen, nitrogen, and water molecules in the upper

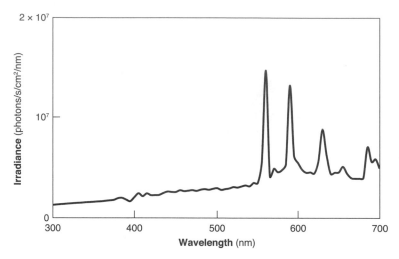

Figure 3.16: Irradiance on a moonless night. Data from Johnsen et al., 2006.

layers of the atmosphere that were split by sunlight during the day. Airglow occurs during the day as well, but is far too dim to see. Because the light comes from specific chemical reactions, rather than general thermal radiation, airglow spectra have many sharp peaks. In the visible range, the sharpest peak is at 558 nm and due to excited oxygen atoms about 90 km to 100 km up in the atmosphere. This peak creates a greenish glow that can be seen from space. The next two brightest peaks are also due to atomic oxygen and found in the red portion of the spectrum. Atomic sodium and rotating and vibrating hydroxyl ions comprise much of the rest, but there are a host of finer features that some physicists have spent their lives studying. On the whole, airglow has low radiance, but it comes from the whole sky, so its contribution to the down-welling irradiance on moonless nights can be significant. Its actual contribution depends on where we are in the solar cycle, airglow being higher during sunspot maxima.

So the irradiance on moonless nights is a sum of these three components—integrated starlight, zodiacal light, and airglow—whose relative contributions depend on the sunspot cycle and where the Milky Way is in the sky (figure 3.16). On average though, it is significantly red-shifted compared to daylight or moonlit nights and massively red-shifted compared to deep twilight. Therefore, for those animals that can see it, the sky goes through some impressive color changes every time the moon or sun rises or sets.

For simplicity, let's just consider a setting sun, though the same principles apply for the rising sun and the rising and setting moon. When the sun is high in the sky, the down-welling irradiance is fairly white to our eyes. As the sun drops below 10°, the irradiance becomes progressively redder. However, just before it reaches the horizon, the irradiance becomes bluer, quickly passing through white on the way. There is a sharp increase in blueness as the sun drops below the horizon, followed by a steady increase during civil twilight. During nautical twilight, the blue intensifies, and moves toward purple as the path length of direct sunlight through the atmosphere increases and ozone absorption plays a larger role. Later, during astronomical twilight, the irradiance becomes redder, due to the influence of airglow and integrated starlight.

This shift from white to red to white to blue to purple to white to red can happen over a period of an hour or two, depending on where you live. It is of course influenced by clouds and other factors, but the bottom line is that, whenever the sun or moon rise or set, the brightness and color of the sky go through some serious gyrations. These likely tax the visual abilities of any animal. It has been argued that the crepuscular periods of dawn and dusk are relatively safe havens from predation (e.g., Yahel et al., 2005). However, others have argued that predation increases during these periods (e.g., Hobson, 1968; Helfman, 1986). This has led to a long and complex literature in which the only point of agreement seems to be that dusk and dawn are special. In my opinion, both hypotheses are likely true, depending on the relative visual abilities of the predators and prey. Closer to home, it has long been known that automobile accidents peak during civil twilight (reviewed by Leibowitz and Owens, 1991).

The Aurora

The only other significant natural source of nocturnal illumination is the aurora. The aurora can in some ways be thought of as the ultimate airglow (Plate 1). As with the latter, the aurora is due to light emission by excited molecules in the upper atmosphere. In fact, the same molecules are involved. Both the green and red auroras are due to emission by excited oxygen atoms. Where the two phenomena differ is in the source of the exciting energy. In the case of airglow, the atoms are excited by solar photons. In the case of the aurora, the atoms are excited by charged particles ejected from the sun.

In addition to emitting a tremendous number of photons, the sun is also spitting out a continuous hail of charged particles collectively known as the "solar wind." These particles, primarily electrons and protons with a few ionized helium atoms, slam into the earth's magnetic field at about one million miles per hour. When the sun is at relatively peaceful point in its sunspot cycle, the vast majority of these particles are guided around the earth by the geomagnetic field, which gets horribly distorted in the process. However, when the sun is particularly active, and especially when it is throwing off large chunks of gas, the solar wind can punch through the earth's field and reach the atmosphere. These particles, while no longer traveling a million miles an hour, still pack a fair punch and stimulate photon emission from oxygen atoms.

The interaction between the solar wind and the earth's magnetic field is such that these particles can usually only break through in regions that are about 2000 miles from a magnetic pole. So, in the northern hemisphere, you usually only see auroras in Alaska, Canada, and Siberia. However, this region expands significantly during large solar storms. For example, during the fairly intense solar storm of September 7, 2002, the aurora could be seen all over the northern United States and photographed as far south as North Carolina.

The aurora ranges from a slight increase in the background radiance to a life-changing sky-wide color show with the irradiance of full moonlight. To my knowledge it is not known whether it has any effect on the circadian rhythms or behavior of polar species. Given the lunar periodicity of so many behaviors and the lack of normal day and night over much of the year in high-latitude habitats, one might guess that occasional intensely bright nights could affect reproductive timing, migrations, and other important behaviors. It would certainly be fun to study.

LIGHT POLLUTION

The final contributor to nocturnal illumination is artificial and sadly becoming increasingly common. The intensity of light pollution of course varies dramatically depending on where you are and the cloud cover (because clouds reflect terrestrial lights), but has become overwhelming over most of the industrialized world (see Garstang, 2004, for an astronomer's view on

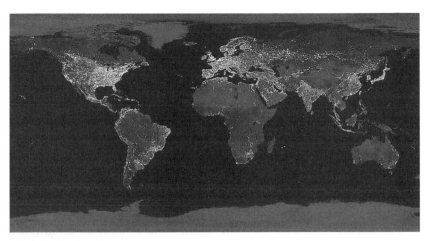

Figure 3.17: The earth at night. Courtesy of the NASA/Goddard Space Flight Center.

the topic). In the United States, the only region east of Dallas, Texas, that is truly dark is the southern half of West Virginia (figure 3.17). The western half of the country has larger dark regions, but not nearly as many as you might expect. In urban areas, the nocturnal sky is generally substantially brighter than a rural sky under the full moon. About 80% of the U.S. population lives in cities, which means that at least four out of five people seldom see much more than the moon, a few planets, and possibly a handful of bright stars. The thought of light traveling billions of years from distant galaxies only to be washed out in the last billionth of a second by the glow from the nearest strip mall depresses me to no end.

My feelings aside, light pollution can wreak havoc on the nocturnal behavior of many species. This has been studied from many perspectives. Some of this work has focused on either the increase in overall irradiance or the effects of local bright sources. For example, light pollution's effect on down-welling irradiance is known to inhibit vertical migration in the freshwater zooplankter *Daphnia magna* (Moore et al., 2000). On land, various small animals forage less under increased illumination, presumably because they are more vulnerable to predation (reviewed by Longcore and Rich, 2004).

A second and larger group of researchers examines the effects of local bright lights. Anyone who walks around on summer nights knows that insects and birds seem to be trapped by artificial lights. Light trapping is found in numerous species, including many aquatic ones, and is thought to be me-

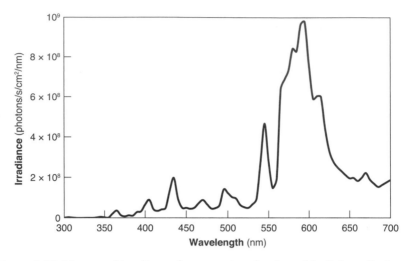

Figure 3.18: Nocturnal irradiance from a region dominated by light pollution (Jamaica Pond, Boston, MA). Data courtesy of Marianne Moore.

diated by several mechanisms. My favorite is that it is solar navigation gone awry. Many species use the sun or moon as a reference point, traveling at a set angle to it. This works well, but only because the sun and moon are far away. Try this with a streetlight and you will go in spirals.

Local bright lights are also known to affect hatchling sea turtles. After emerging, turtles find the ocean by moving away from the dark dunes and toward the relatively bright ocean. The lights of beach houses reverse this gradient and can cause the animals to wander off into the dunes and die. Sea turtles are apparently cuter than insects and birds, because, once this link was established, lighting ordinances were quickly established for beaches with nesting turtles.

Researchers have also looked at the effects of artificial lighting on communication, competition, reproduction, pollination, and a host of other behaviors. To my knowledge, though, none of these studies examine the effect of the color of light pollution. While humans use a number of different artificial light sources, light pollution is predominantly due the mercury-vapor lamps and low- and high-pressure sodium lights that light streets, public places, and commercial and industrial real estate. As such, light pollution tends to have a particular spectrum quite different from that of any natural illumination. First of all, it is orange, having a color temperature of about 2700K (figure 3.18). It is also highly saturated, with almost all the photons in a narrow wavelength band centered on 590 nm (the location of the two "sodium-D" lines). The relative number of photons in the blue region of the spectrum is small. Now that we know certain animals can see color at night, and use it to find flowers, the odd spectrum of light pollution needs to be taken into account. But enough of this depressing subject. On to bioluminescence.

BIOLUMINESCENCE

As I mentioned at the beginning of the chapter, bioluminescence is, along with the sun, one of the two major sources of light on the planet. Yes, there are the aurora, airglow, starlight, and even some lightning now and then, but, for the vast majority of organisms, nearly all the photons they see, or otherwise interact with, come from either the sun or a bioluminescent organism. The reason we don't pay much attention to bioluminescence is because over 80% of the light-producing genera (and the vast majority of the light-producing individuals) are marine. For unknown reasons, bioluminescence is rare on land, limited to fireflies, glow worms, and a grab bag of random arthropods, mushrooms, and worms (Herring, 1987, has the most complete list of bioluminescent taxa). Freshwater bioluminescence is even rarer, which we will return to later. However, in marine environments, and particularly in the water column of the open ocean, bioluminescence is ridiculously common. It has been stated that at least 90% of oceanic life emits light. While statements of this sort are always a bit silly (90% of animals, of species, of biomass? How could you ever count?), any research cruise to the open sea will quickly convince you that bioluminescence is as common as dirt. Look over the side of the ship at night and the wake glows. Go for a night scuba dive and every move you make is punctuated by blue flashes. Flush the toilet and you are in for a light show (many research ships use saltwater plumbing to save water, so the toilet bowls are full of bioluminescent organisms). Overall, bioluminescence is so common that a friend of mine once published a paper whose sole conclusion was that a certain species of an otherwise highly bioluminescent group did not in fact emit light (Haddock and Case, 1995). Bioluminescence is enzyme-mediated chemiluminescence. Before we get into the details of this, let's just stop and appreciate the fact that nonthermal sources of light emission exist. As we mentioned earlier, all light is created by moving charged particles. At the molecular level, everything is always moving, so matter is always radiating. However, as we also discussed, the spectrum of this radiation does not have a significant component in the visible until things get to at least 650°K. Even if any animal could survive at that temperature, the black body spectrum is so broad that it is a terribly inefficient way to produce visible light. For example, the luminous efficiency (perceived brightness per Watt) of an incandescent 60-watt lightbulb is only 2% (that of a 555 nm green laser pointer is 100%). The luminous efficiency of our roasting 650°K animal is so low that it is not even worth mentioning.

Suppose we discovered an alien life form that produced light by somehow surviving a body temperature of 6600°K. An object at this temperature has the highest luminous efficiency possible for a black body radiator—about 14%. This sounds bad, but is actually pretty good, about equal to a high-quality fluorescent bulb. Unfortunately, the alien would also be emitting appalling amounts of radiation at longer and shorter wavelengths. Remember that, ignoring losses in the atmosphere, the irradiance you get from the sun depends only on its temperature and angular area. So, if the alien were the size of a penny (1 cm diameter), standing 114 cm from it would feel like standing outside on a sunny day—with one big exception. There would be no ozone between you and the alien to absorb the short-wavelength ultraviolet radiation. You would feel nice and toasty warm, and then die slowly from radiation damage. If the alien were anything close to human-sized, it would have to stay at least 100 meters away or you wouldn't need to wait for the ultraviolet radiation damage. You would roast alive. Just like the Greek myths always told you, being next to a god who is as bright as the sun will kill you, assuming that gods are black body radiators. The upshot of all this is that, even if we make great allowances for what is possible, thermal radiation simply cannot be used for visual signaling in animals. There are, of course, animals that can detect the thermal radiation of objects at physiologically reasonable temperatures, but only in the infrared.

The reason black body radiation is so inefficient is that it's a nonspecific process. You heat something up, particles move around and radiate, but you have no control over what wavelengths of radiation you will get. It's like shaking a box of Legos and hoping that you get a car. Light produced by animals needs to be made to order, and chemiluminescence provides that selectivity.

Rather than relying on the motion of all charged particles, chemilumines-cence depends on changes in the motions of a small and specially chosen subset. There are a number of ways of interpreting what happens, running the whole gamut from classical to semiclassical to quantum optics. The mul-tiple interpretations can get confusing, so I've chosen incompleteness over confusion and will just describe this in one hopefully intuitive way. I would like to spend a little time with this, because the same process will show up in reverse when we discuss absorption in chapter 4.

Imagine a hydrogen atom with its single proton and electron. The elec-tron is said to orbit around the proton. You can imagine this as one ball re-volving around another. In reality, it is nothing like this and possibly not even describable, but what you want here is a mnemonic model, not a true description of the universe. If you took chemistry, you can imagine those bi-zarrely shaped orbital clouds. Whatever works for you.

This atom has a certain internal energy, which depends on the relation-ship between the electron and the proton (we are ignoring the kinetic energy of the atom as it moves through space). Let's assume that the atom is in its ground state, which means that the internal energy is as low as it can be. Now suppose that some process (the nature of which is unimportant) works on the atom, resulting in an increase in the internal energy of the atom. You can think of the electron moving to a higher orbit, changing the shape of its orbit, or dancing a little Irish jig. It matters little, and you will never know what it looks like anyway.

What does matter is twofold. First, the internal energy of the atom can only jump to certain levels. This is one of the central tenets of quantum me-chanics. The levels, often depicted as a vertical stack of parallel lines, are well known for hydrogen and a few other simple atoms. Beyond this, things get complicated fast, and analytical solutions are rare. However, the energy lev-els remain discrete even for complex molecules, though they can get so close together that they appear continuous.

Second, a short time later, the energy level of the atom/molecule will drop back to its ground state. When this happens, a photon will be emitted (figure 3.19). The energy of this photon will equal the energy of the drop, which usually equals the energy of the original jump (sometimes the trip down to the ground state happens in multiple hops, which we will get to in chapter 7). Remember from chapter 2, that the energy of a photon equals $h\nu$, where h is Planck's constant and ν is the frequency. So the frequency of

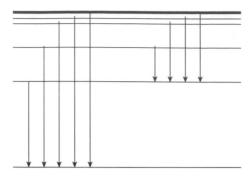

Figure 3.19: Various possible energy-level drops in a hydrogen atom. The longer the drop, the greater the energy and the shorter the wavelength of the emitted photon.

the emitted photon equals E/h, where E is the size of the energy drop. This means that its wavelength equals $c/v = ch/E$, where c is the speed of light in the medium in which the photon is emitted. In other words, find a way to add a certain amount of energy to an atom/molecule, and you can make almost any wavelength of light you want.

Bioluminescent organisms take advantage of this process to produce visible light. In their case, the excitation is provided by a chemical reaction. While the actual reactants vary by taxonomic group, they all involve the oxidation of a small and usually dietarily derived molecule, known as a "luciferin." The oxidation is mediated by an enzyme, known as a "luciferase," and generally requires energy via ATP. The taxonomic distribution of the luciferins is actually fairly perplexing, as is the evolution of bioluminescence itself. An excellent recent review of these issues can be found in Haddock et al. (2010). The reaction itself occurs in one of three places. In many fish and a few cephalopods, the light is produced by symbiotic bacteria that live in a pouch. The bacteria produce light continually, so the animal uses a shutter to control whether the light actually exits the body. The symbiotic relationship between the bacteria and the host is quite deep and has been explored in detail by Margaret McFall-Ngai (reviewed by Nyholm and McFall-Ngai, 2004). In many other cases, the light is produced by the animal itself inside cells known as "photocytes." These cells are often part of beautifully complex organs that contain filters, mirrors, lenses, and other apparatus for controlling the color and direction of the light. Finally, some marine species perform the chemical reaction outside of their bodies, essentially vomiting the

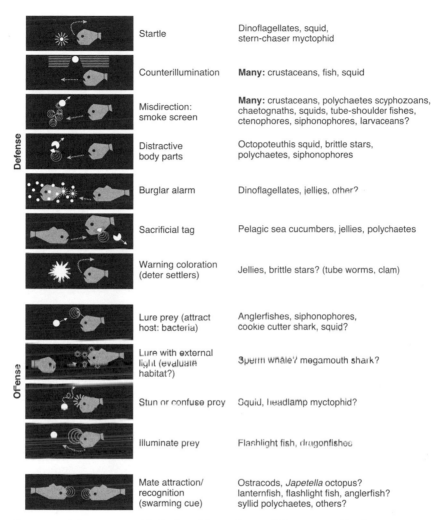

Figure 3.20: Various established and hypothesized functions of bioluminescence in marine species. From Haddock et al., 2010.

necessary reactants into water. Certain shrimp are masters of this, producing brilliant clouds of what is termed "spew bioluminescence."

Despite the ubiquity of bioluminescence, its functions are poorly understood, primarily because it mostly occurs in open-ocean species that are difficult to collect and whose behavior is nearly impossible to study. A host of functions have been proposed (figure 3.20). However, with the exception of luring, counterillumination, and communication (for fireflies), most of these proposed functions are still in the hypothesis stage. Even the case for luring is primarily based on the fact that a glowing ball hanging in front of a toothy mouth could hardly be anything else but a lure. As far as I know, there is no direct evidence of the lure actually working. This leaves sexual signaling in fireflies, which has been well studied (research is so much easier on land), and counterillumination, which can be observed directly in the lab. Counterillumination is indeed wonderful and gets around the problem that, no matter how white you make your ventral surface, you'll still appear as a black silhouette when viewed from below (Johnsen, 2002). This is because the down-welling light in the ocean is orders of magnitude brighter than the up-welling light, which means that, when viewed from below, the light illuminating a surface is far less bright than the background light. Since a surface can only reflect up to 100% of the light, it always looks nearly black when viewed from below.

Thus, many animals cover their ventral surfaces with photophores and make sure that they match the intensity of the down-welling light, even keeping up with passing clouds. Some species of squid even modulate the color of the light to better match the down-welling light. This is a good trick, because cephalopods (with one exception) appear to be color-blind (see Mathger et al., 2006, for what may be the last word on this contentious topic). In at least some cases, they seem to use water temperature as a proxy for water color, going on the usually accurate assumption that warmer water is shallower and greener (Herring et al., 1992). So if you put these animals in warmer water, their ventral surface glows green. Put them in colder water and they glow blue.

As you might guess, an entire book could be written about the chemistry, evolution, and functions of bioluminescence, so I have to regretfully confine myself to a few optical aspects and puzzles. First, bioluminescence is usually quite dim, ranging from 10^3 photons/s for a single bacterium to 10^{12} photons/s for some krill and fish (reviewed by Widder, 2010). This latter number might seem fairly large but remember that it is for all wavelengths combined and for light emitted in all directions. While the brighter flashes can be seen under moderate room light, most bioluminescence can only be seen by the dark-adapted eye. While bioluminescence is a far more efficient way of producing light than thermal emission (at least for organisms), producing enough light to be easily seen under daylight conditions is expensive. The central problem is not energy expenditure, but the fact that most luciferins are derived from diet and can be used only once. We see this a lot at sea when we collect animals using a trawl net. As you might guess, life in a trawl net is no better than life in the scoop of a bulldozer being pushed down a busy sidewalk. Most bioluminescent species flash like crazy after being caught and have used up nearly all their luciferins by the time we get them on board. If we are interested in studying the bioluminescence of animals, we find nicer ways to capture them.

Second, despite the fact that chemical reactions can provide energies over a wide range, bioluminescent spectra are not that diverse. They generally have a Gaussian shape (i.e., they look like a bell curve) and so can be pretty well described if you know the peak wavelength and the width of the curve (the width usually given as the range over which the spectrum is at least half the value of the peak value). The peak wavelengths also don't vary by much, generally from about 450 nm to 550 nm, with most of them falling near 480 nm. In other words, bioluminescence, with rare exceptions, is blue or green (figure 3.21). There are a few cases of yellow bioluminescence and some that are vaguely purple. Red bioluminescence is extremely rare, the only cases being railroad worms (larvae of the beetle *Phrixothrix* sp.), which have red photophores on their heads; the deep-sea siphonophore *Erenna* sp.; and three species of deep-sea fish, which will be discussed in the next chapter. With the exception of the railroad worm and possibly *Erenna* sp., the light created by the oxidation of the luciferin is not red, but is converted to a longer wavelength via fluorescence and/or long-wavelength filters.

One could argue that the predominately blue and green nature of bioluminescence is driven by ecology. After all, most light-producing species are marine, and ocean water transmits blue and green light the best. The fact that oceanic bioluminescence is usually blue, and coastal bioluminescence is usually green, strengthens this argument since oceanic and coastal waters optimally transmit blue and green, respectively. However, I suspect that this isn't the whole story. First of all, terrestrial bioluminescence is usually green. This could be due to the fact that the primary viewers of terrestrial bioluminescence—insects—are quite sensitive to green light, but this seems a bit hand-wavy to me. Also, the fact that nearly all red bioluminescence starts out at a shorter wavelength suggests that some sort of biochemical constraint is at work. After all, the fluorescence and filtering required to convert green light to red is quite inefficient. The filtering in particular removes most of the emitted light. There also appears to be no ultraviolet bioluminescence—

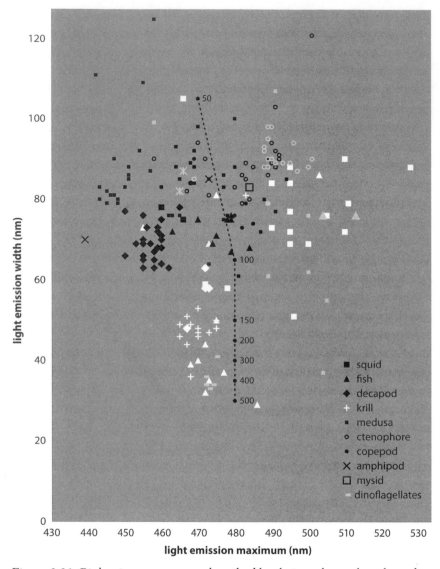

Figure 3.21: Bioluminescent spectra described by their peak wavelengths and the widths of their emission curves. White symbols denote bioluminescence that is used for counterillumination. Gray symbols denote shallow species. Dotted line shows the peak wavelength and spectral width of the down-welling light as a function of depth. From Johnsen et al., 2004.

though of course finding it would be a challenge. Interestingly, just as the protein component of our visual pigments controls what wavelengths of light are most likely to be absorbed (more about this in the next chapter), the luciferase appears to control what wavelengths of light are emitted when luciferin is oxidized. We know that red bioluminescence of the railroad worm depends on its luciferase, but not why luciferases of this sort are so rare. Alison Sweeney once proposed an optical constraint model that more or less went like this: (1) blue light is transmitted best in the ocean, so (2) most deep-sea animals only see blue light well, so (3) bioluminescence is evolutionarily constrained to be blue (the same argument works for green coastal waters). A prediction of this model is that, once you get below 1000 meters and run out of solar light, this constraint may relax and you should see bioluminescence in more colors, and visual pigments of greater diversity. Indeed, one of the few examples of red bioluminescence, in the siphonophore *Erenna*, was found at depths of 1600–2300 m (Haddock et al., 2005). However, in this case it looks like the red light is due to the conversion of shorter wavelength light via fluorescence.

Another odd fact about bioluminescence is that, while it is ubiquitous in the ocean and fairly common on land (found in many insects, myriapods, fungi, the gastropod *Quantula striata*, and even some earthworms), it is extremely rare in freshwater. The only bioluminescent organisms in freshwater are the larvae of some insects and the limpet-like snail *Latia nerotoides*. This snail is found only in New Zealand streams and produces a green light via a luciferase-luciferin reaction that uses a unique luciferin (Ohmiya et al., 2005).

Two main reasons are given for the rarity of freshwater bioluminescence, neither of which I find terribly convincing. The first hypothesis is that freshwater generally doesn't provide a useful optical environment for bioluminescence because it is not deep and is often turbid. The depth argument makes no sense to me. Even excluding the fact that some lakes are extremely deep, depth is only required for darkness, which is easily supplied by the night. After all, there are plenty of terrestrial bioluminescent species. The relative murkiness of much freshwater makes more sense, since a dim signal won't travel far in turbid water. However, there are a fair number of lakes and streams that are at least as clear as coral reef waters and some that are as clear as blue oceanic waters. The other major hypothesis is that freshwater systems aren't stable enough for bioluminescence to evolve. While it is true that many streams and lakes can be quite ephemeral, even on human time scales, some are ancient, particularly the rift lakes. Lake Baikal, for example, is over 25 million years old and incidentally also clear and quite deep (average depth 750 m). It has had time to evolve 1000 endemic species, including truly odd creatures like the transparent *Golomyanka*. Also, even though many freshwater systems have short lives, freshwater as a whole is ancient and has a set of species that are found in many locations. Additionally, it has been estimated that bioluminescence has evolved at least forty times (Haddock et al., 2010), so the "not enough time" argument isn't as convincing. Finally, this is not a testable hypothesis.

I'm not sure what's going on. It's hard to develop a biochemical or osmotic argument based on the differences between fresh- and saltwater, given that bioluminescence is often intracellular. Perhaps, given that luciferins are often derived from diet, they are relatively rare in freshwater habitats and so the required substrate simply isn't available. I don't know, but would love an answer before I die.

Two final remarks before I end this all-too-brief review. First, despite common usage, phosphorescence and fluorescence are not synonymous with bioluminescence. The first two processes convert existing photons to those of a different energy. They cannot make light de novo. Most luciferins are fluorescent, and many bioluminescent organisms make use of fluorescent molecules to alter the spectrum of the emitted light (the famous green fluorescent protein—GFP—being the best known example), but ultimately bioluminescence makes light from a chemical reaction.

Second and finally, bioluminescence is beautiful. I have no idea why small blue and green lights on a dark background have such an effect on people, but they do. Swimming in a cloud of bioluminescent dinoflagellates at night or leaving glowing footprints on the beach is a transcendent experience. Even if you never go on an ocean research cruise, or flush a saltwater toilet, make a point to include some bioluminescence in your life.

Mechanoluminescence

In theory, any form of energy can be converted into light. Ultimately, light is always due to the acceleration of charged particles, but various processes can get you to that point. In thermal radiation, the acceleration of the charges is directly related to their temperature. In chemiluminescence (e.g., airglow, aurora, and bioluminescence) the acceleration is the result of specific chemical reactions. But what about mechanical forces? Can you bend something and make it glow?

It turns out that you can. Certain materials emit light when you deform them. This is known as "piezoluminescence." I am not expert on the subject, but it appears to be relatively uncommon and not that dramatic. More common and potentially biologically relevant is triboluminescence, which is light emission caused by breaking something. Stirred sugar and crunched candy show this effect, but, since I have been a little kid, there has always been one classic example of this—the Wint-O-Green LifeSaver (Plate 2). If you turn out the lights and crunch on one of these in front of a mirror, you'll see bright flashes of blue light. Linda Sweeting has spent a portion of her career studying this effect and the following explanation is based on her wonderful and accessible article "Light in Your Candy" (Sweeting, 1990).

When you bite candy of any kind, you crack sugar crystals, which can separate charges. In other words, more electrons end up on one side of the break than on the other. If the charge imbalance is great enough, the electrons will shoot across the gap at high speed. As they do this, they smash into the intervening air and excite nitrogen molecules, which then emit light. In other words, you've created a miniature lightning bolt in your mouth. Sounds crazy, but Sweeting measured the spectrum of light emitted by crushed sugar and found that it matches that of lightning (who knew?).

The problem is that the emission spectrum of nitrogen is mostly in the ultraviolet. Part of the emission is in the visible range, but it's fairly dim. As impressive as lightning is to us, imagine what it looks like to animals with ultraviolet vision. However, to our limited sight, the light emitted by your average candy isn't so exciting and you can stir sugar like a demon and not be that impressed. I can tell you this from personal experience.

Wint-O-Greens are much brighter because they are fluorescent. We will discuss fluorescence more in chapter 7, but for now just accept that fluorescent materials absorb light of one wavelength and emit light of a different, generally longer, wavelength. Oil of wintergreen is particularly good at absorbing ultraviolet light and re-emitting visible light. The emitted light has a broad spectrum, but appears blue to our eyes. So a mouthful of fluorescent lightning is only a short walk to the store away.

Triboluminescence shows up in other places too. Attach a long strip of duct tape to a table and then yank it off in the dark. You'll see flashes of light right where the tape is leaving the surface of the table (Camara et al., 2008). Quartz crystals also emit light when broken or struck together (Walton, 1977, has a long list of triboluminescent substances and their history). There is a persistent on-line story that the Uncompahgre Ute Indians of central Colorado built ceremonial rattles of translucent rawhide and filled them with quartz crystals. When shaken they produced a dim light that could be seen in the dark. There are pictures of the rattles on the Web, and quartz is indeed triboluminescent, so it's possible. I've found no official documentation of this, but I did buy crystals of quartz from the local museum shop and banged them together in my darkened bathroom. The light they produced was brighter than I expected. Whether it is triboluminescence remains to be seen.

So are there any biological implications of triboluminescence? Perhaps. Deep-sea hydrothermal vent communities are some of the strangest ecosystems on earth. The local residents not only have to deal with the pressures found miles under water, but also temperatures that range from 4°C to 350°C, sometimes over distances of only a meter. Add the toxic chemicals belched out by the volcanic vents, and you can see why most of the deep-sea vent sites have names like Snake-Pit and Hole-to-Hell.

At these depths, light from the sun has long since been absorbed by the intervening water. I'm sure the stray photon or two makes it down there, but it's not enough to see by. Bioluminescence has not been a hot topic of research at deep-sea vents, and seems to be relatively rare. Like everything, the vents radiate, but the visible radiation of a black body at 350°C (= 623°K) is quite low and predominately red. So researchers were surprised to find that a common inhabitant of Atlantic vent communities, the shrimp *Rimicaris exoculata*, has eyes. The eyes are unusual and were at first missed, which explains the species name. Instead of the usual camera eye or compound eye design, each eye has a large flat surface with no imaging optics, apparently designed for one purpose—to detect dim light (Van Dover et al., 1989).

Despite their nasty conditions, hydrothermal vents are a tremendous source of food for bacteria, which in turn are the basis for a thriving ecosystem. Each community is essentially an oasis surrounded by a low-calorie abyssal plain. Therefore, the obvious hypothesis was that the eyes of *Rimicaris* were used to help the animal find the communities via their thermal emissions. However, subsequent research showed that the visual pigment extracted from the animals' eyes was most sensitive to blue and green light, making it a poor match for finding black body radiation from the vents. This led to measurements of the emission spectra of the vents themselves (*not* easy to do), which showed that they emitted substantially more visible light than would be expected by black body radiation—nineteen times more in the red part of the spectrum (Van Dover et al., 1996). These measurements were followed by others (White et al., 2002) who found that the difference between the measured light and that predicted by a black body was even greater at shorter wavelengths. For example, at 500 nm, there was 10,000 times more light than could be explained by the temperature of the vents.

The source of this extra light is still unknown, but triboluminescence is on the short list. If you have ever poured boiling water into an ordinary glass, you know that sudden temperature changes can make rocks crack. While vents are not continually vomiting lava, they have large temperature gradients that stress and crack the rocks. In addition, cooling magma crystallizes, which can lead to a sister source of light emission called "crystalloluminescence." Chemiluminescence and bioluminescence are also possibilities. Additionally, it is unknown whether *Rimicaris* is paying attention to vent light at all. That said, I personally think it would be wonderful if deep-sea animals were finding volcanoes by the light of cracking rocks.

Sonoluminescence

The last source of light emission seems so esoteric that you might imagine it has no biological connection. However, in the last decade, it has been found in two groups of animals, proving once again that nearly every physical process exists in the biological world, somewhere.

Sonoluminescence is light produced by collapsing bubbles of air in water. It was originally discovered by Frenzel and Schultes (1934), two sonar workers who put an ultrasound speaker in a tank of film developer to see if it would shorten processing time. Instead, they found little spots all over their prints that they realized were due to light emitted by collapsing air bubbles.

The study of this process has been refined over the years and now the sonoluminescence of single bubbles can be studied in detail. Despite this concerted effort, there is no real agreement about what causes the light emission. It is known that the bubbles collapse exceedingly quickly. The radius shrinks at supersonic speeds, creating a massive, but short-lived increase in pressure. This pressure increase leads to an equally massive and short-lived increase in temperature. The original estimates for this temperature were as high as one hundred million degrees, but have been revised downward to between 6000°K and 20,000°K (see excellent review by Brenner et al., 2002). This is still extremely hot, so a primary hypothesis is that sonoluminescence is thermal radiation. The dust has yet to settle though.

The link between sonoluminescence and biology is cavitation. Water is difficult to compress or stretch. If you fill a sealed syringe with water, you will have a hard time yanking up on the plunger. However, if you yank hard enough, you will create an air bubble in the water. I suppose in a way you could say that you have torn the water. This process, called "cavitation," occurs whenever the pressure in the water drops below water's vapor pressure. The latter value depends on temperature—at room temperature it is about 0.02 atmospheres. This is low, but achievable. It happens all the time with boat propellers.

Once the local pressure rises, the bubble quickly collapses. The collapse is violent enough that it creates shockwaves that can punch holes in metal. Manufacturers of boat propellers spend a lot of time worrying about this. The collapse can also produce sonoluminescence.

At least two aquatic animals move quickly enough underwater to cause cavitation. The first discovered to do this were snapping shrimp (Lohse et al., 2001). These odd creatures look like little lobsters with asymmetric claws. The larger claw has a cocking mechanism that allows it to build up a significant amount of tension in one muscle that is then released by a second muscle-activated trigger. This creates a high-speed jet of water that leaves a cavitation bubble in its wake. The collapse of this bubble creates an impressively loud snapping sound (190–210 dB re 1 μPa at 1 m) that is used for territorial defense and can be heard by the thousands on some coral reefs. Lohse et al. (2001) studied this cavitation and found that it resulted in sonoluminescence. The flash lasts less than 10 nanoseconds and only produces a total of 5×10^4 photons, so it may be below the visual threshold of all animals.

The other cavitating animals are mantis shrimps (stomatopods), a group of crustaceans famous for many optical tricks. They are known to express at least twelve different visual pigments, giving them perhaps the most complex color vision known. They can also see both linear and circular polarized light and have patterns on their bodies that can only be seen by animals with these abilities (more about this in chapter 8). They also make some seriously weird noises. However, the relevant point here is that they have a truly nasty pair of appendages that they use to build burrows and kill prey and conspecifics. These appendages are modifications of the second thoracic legs and come in two flavors—spearers and smashers. The spearers have a long sharp tip and the smashers have a club. The smashers can be used to break shells and can strike with a force of up to 1500 N over an area of just a few square mm. This is equivalent to a weight of 336 lbs (imagine balancing an unsharpened pencil on your stomach and then having someone put a washing machine on top of it). This force is enough to cavitate water, which has been proven using high speed photography (Patek and Caldwell, 2005). It is likely that this cavitation also results in sonoluminescence, but this has not been documented.

There are a few other natural sources of light (e.g., lightning, lava, methane explosions), but for the sake of brevity, it is probably best to now move on to what happens to light after it is emitted. This is covered in the next four chapters on absorption, scattering, and fluorescence.

Further Reading

Much of the early part of this chapter is based on the first chapter of Bohren and Clothiaux's *Fundamentals of Atmospheric Radiation*, which has a nice discussion of black body radiation that clears up many misunderstandings The first chapters of John Lythgoe's *Ecology of Vision* review natural illumination. Unfortunately, the book, published in 1979, can be hard to find and is expensive.

The *RCA Electro-Optics Handbook*, published by Burle Industries, is a surprisingly good source of information about black body radiation and natural and artificial light sources. It also has clear chapters on unit conversions. Best of all, the older editions are short and compact, and therefore easy to take into the field. My favorite is the 1974 edition.

Two good books (in fact, the only two books I know of) on sunset and twilight are Aden and Marjorie Meinel's *Sunsets, Twilights, and Evening Skies* and Georgii Rozenberg's *Twilight: A Study in Atmospheric Optics*. They are both fairly technical, especially the latter, but have excellent information that can be found nowhere else.

For a more aesthetic approach, try *Color and Light in Nature*, by David Lynch and William Livingston. This book is a complete rewrite of Minnaert's famous *The Nature of Light and Colour in the Open Air* and has excellent, explanatory diagrams and beautiful color photographs of just about every atmospheric phenomenon you can think of.

E. Newton Harvey's *A History of Luminescence from the Earliest Times Until 1900* is exactly that—a comprehensive history of the phenomenology of nonthermal radiation. Goes all the way back to the Greeks and covers substances you have probably never heard of. You have to be careful though, because, while he knows the difference between phosphorescence and bioluminescence, some of the sources he quotes use the terms interchangeably.

The Astronomical Applications Department of the U.S. Naval Observatory maintains a website that allows you to calculate the position of any astronomical body as a function of time, date, and geographic location. It is particularly useful if you are interested in twilight or nocturnal work. The current Web address is http://aa.usno.navy.mil.

CHAPTER FOUR

Absorption

You can't have a light, without a dark to stick it in.
—ARLO GUTHRIE (from the album *Precious Friend* by Guthrie
and Pete Seeger)

Absorption has been called the "death of photons" (Bohren and Clothiaux, 2006). While the energy of a photon is never truly lost (reincarnation is a fact in physics), most people find the conversion of photons into heat and chemical reactions less appealing than their original emission. I admit that I enjoy watching bioluminescent plankton more than contemplating the blackness of my T-shirt.

Without absorption, though, the earth would be a far less colorful place, with no paintings, flowers, leopard spots, or stained-glass windows. There would still be some colors due to interference and scattering (to be discussed in chapters 5 and 6), but we would be blind to them because our vision is based on absorption. More importantly, if nothing on earth absorbed light, the planet would be lifeless and cold.

Also, while we tend to think of absorption, scattering, and fluorescence as separate phenomena, at heart they are sides of the same coin. Therefore, understanding absorption helps us understand all the interactions of light with matter. So, on to the dark.

LIGHT AND MATTER

When you get down to it, a photon can do only two things: it can bop along on its merry way through space or it can interact with matter. The former may be boring or it may be fascinating. We have no way of knowing. According to some interpretations of quantum mechanics, what a photon does in transit is unknowable, even in principle. The only thing we can actually observe is light's interaction with matter, or—more specifically—with atomic nuclei and electrons.

As with everything in light, there are two ways to look at this interaction, in this case called "classical" and "quantum-mechanical." I cannot possibly do justice to each here, but hope to give you a sense of both.

In the classical view, light is an electromagnetic wave that is both produced by the acceleration of charged particles and causes their acceleration in turn. This is easier to think about if you put visible light aside for a moment and start with radio waves. My local head-smashing classic rock station is at 100.7 on the dial. This means that the waves produced by this station have a frequency of about 100.7 million cycles per second (megahertz or MHz). I say "about" because the frequency is modulated to carry the actual music, which is why it is called "FM" (frequency modulation) radio. This modulation is a tiny fraction of the carrier frequency, though, so we will pretend it's all at 100.7 MHz. The wavelength of any wave is just its velocity divided by its frequency, and the velocity of any electromagnetic wave in air is close to 300 million meters per second (we will get more into what the speed of light actually means in chapter 5).

Therefore, the wavelengths of the waves from my station are close to three meters long (= 300,000,000/100,000,000). The waves from AM radio are even longer. My hometown station KDKA broadcasts at about one MHz, and so produces waves that are three hundred meters long. As an aside, you can see why radio engineers prefer not to use photons. It's hard to reconcile a microscopic packet of energy with a 300-meter-long wave. As a further aside, the antennas for radio transmitters and receivers work best when they are on the order of the wavelength in length. This explains why AM transmitter towers are huge and why cell phones operate at high frequencies (~900 MHz).

As the wave from my station passes my car, the electric field in my antenna oscillates at 100.7 MHz. Radio stations produce waves with vertical electrical fields, and my car antenna is also roughly vertical. This means that, over 100 million times a second, the electric field in my antenna goes back and forth between pointing up and pointing down. This seems awfully fast, but it's still slow enough for the electrons in the metal of the antenna to get accelerated by this field. The net movement of the electrons is small, especially when compared to their far higher random thermal motions, but it is significant. These accelerating electrons in turn radiate their own electromagnetic wave, which will have the same frequency as the incident wave—100.7 Mhz. There is nothing special about my antenna. Every metal object you own (including your metal dental fillings) is busily interacting with radio waves.

Of course, it is more complicated than this because my station isn't the only one out there. In Chapel Hill, there are dozens of other FM and AM radio stations, not to mention waves at many other frequencies (broadcast and satellite television, shortwave radio, cell phones, CB radio, navigation beacons, etc., etc.). I've seen allocation charts of the radio band for the United States alone and they look like the subsurface wiring diagrams for Manhattan. Even ignoring everything more energetic than microwaves, the electrons in the antenna are oscillating in response to an extremely complicated electric field.

So how do you ever hear the music? First of all, the length of an antenna affects how efficiently it interacts with an electromagnetic wave. Antenna design is a complicated field, but a rough rule for a simple sticklike car antenna is that it most efficiently interacts with radio waves with wavelengths that are four times its length. This is why car antennas are about 3/4 of a meter long—to best interact with 3-meter-long FM radio waves. You have probably noticed that you don't have a 75-meter-long antenna on your car, but can receive those 300 meter-long AM radio waves just fine. This is because AM antennas often use a different design, and because many local AM stations are ferociously powerful.

So my antenna's length means that it best interacts with the right region of the EM spectrum, but it is not nearly selective enough. The electrons in the antenna would still be oscillating in response to many signals of different frequencies and emitting a complex radio wave. The car's radio is able to untangle this mess for two reasons. First of all, the electromagnetic wave produced by the intersection of a host of other waves is just the sum of the individual waves. Unlike the ingredients in a good soup, waves don't lose their identity when they interact (figure 4.1). You can always separate them again. So, even though the electrons in the antenna are moving in a complex way in response to a complex collection of incident radio waves, you can look at what's happening at one frequency and ignore the rest. We will talk about this more in chapter 6—for now, just appreciate how convenient this is.

But how does the radio know which frequency to pay attention to? This is where resonance comes in. If you are a practicing biologist, you almost

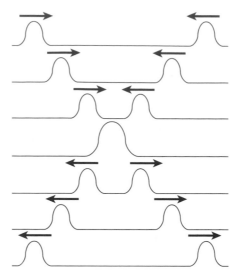

Figure 4.1: Two waves passing each other, demonstrating the superposition principle. When they intersect, the resultant wave is just the sum of the two waves. Also, they are not affected by having passed through each other.

certainly took a semester of introductory physics in college that almost certainly covered harmonic oscillators. You also may have hated that part of the course. After a brief sentence or two about how harmonic oscillators and resonance explain the life of swinging children, many students are thrown to the wolves of critical damping, Q-factors, and dash pots. I have never met anyone who owned or even saw a dash pot.

This is a shame, because oscillation and resonance show up everywhere in biology. Running, walking, flying, and swimming are all examples of harmonic oscillation, and many animals take advantage of resonance to improve the efficiency of their locomotion. Indeed, much of biomechanics is the study of harmonic oscillators. But it doesn't stop there. Nearly every aspect of organismal physiology is cyclical, from heartbeats to circadian rhythms to seasonal migrations to breathing to the cell cycle itself. My college roommate stuck a quote on our door by the philosopher Willard Quine, "to be is to be the value of a bound variable." Not bad, but I would prefer "to be is to be the oscillating value of a bound variable." In short, life is oscillation.

Most things that oscillate have particular frequencies at which they prefer to cycle. Anyone who walks with a partner that is half or double their height is well aware of this, because human legs work like pendulums and have a

Figure 4.2: Resonance in the office. From Randall Munroe, xkcd.com.

natural period that depends on their length. Simple mechanical oscillators, like a clock pendulum or sloshing water in a bathtub, have one resonant frequency (figure 4.2). More complicated oscillators have multiple resonant frequencies. When I was a kid, my dad would entertain me on car trips by driving at various speeds that made different parts of the car resonate and vibrate. At 55 miles per hour, which was unfortunately the speed limit back then, our Dodge Dart had a particularly impressive resonance that made the front seat go into violent shudders.

By now, you may be wondering how a chapter that began with light absorption moved first to FM radio and then to 1970s automobiles. This is because, in the classical view, absorption is resonance. When you drive an oscillator at one of its resonant frequencies—for example, sloshing water in the tub—you are doing work on it in the most efficient way possible. Another way of saying this is that you are efficiently transferring energy to the oscillator, which means it has more and you have less. In other words, your energy has been absorbed.

To be honest, this is a sloppy way of putting it. Despite the fact that we often talk about hot objects as containing heat energy and photons as containing electromagnetic energy, energy is not an object. Despite its equivalence with mass (*a la* E = mc²), energy is better thought of as a concept, sort of a bookkeeping principle that helps you keep track of what will happen in a given situation. For example, I once tossed a refrigerator out of my third-story window. It was going pretty fast by the time it hit the ground. We say that its gravitational potential energy was converted to kinetic energy. But was some mystical potential energy substance actually converted into refrigerator-speed energy? I doubt it, but using the concept of energy and its con-

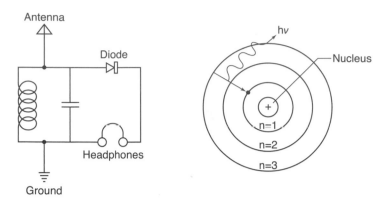

Figure 4.3: (Left) Oscillator for 100 MHz waves (a radio). (Right) Oscillator for 100 terahertz waves (a hydrogen atom). Both are highly simplified representations.

servation helped me predict how hard it would hit the grass in my backyard. Similarly, warm oceans do not contain heat as an incarnate substance. More controversially, photons do not strictly contain electromagnetic energy. This all gets philosophical and subtle quite quickly though, so it is often simplest to go with the sloppy approach and just say that energy is absorbed.

So, back to the car radio. Radios receivers (at least the older simpler ones) are essentially oscillating electrical circuits (the rest of the radio amplifies the signal). In this case the oscillation is electrical, and the resonant frequency is controlled by the tuning knob (figure 4.3). The oscillating circuit is attached to and electrically driven by the electric field in the antenna. So when you set the knob on your radio to 100.7 FM, you are changing the resonant frequency of the receiver circuit to 100.7 Mhz. When this happens, any electrical fields at the surface of the antenna that oscillate at this frequency will create large oscillations in the radio circuit and you will hear music. Energy will be transferred from the wave to the radio and eventually to your ear. Some local stations are actually powerful enough that you can listen to them with a radio that has no power source. As a kid, I used to make a radio out of ten feet of wire, a diode, and an earphone. I could hear KDKA just fine, even without any power source. The clothes dryer belonging to a former student's grandmother picks up the local classic rock station in Rockford, Illinois (you have to do a load of laundry to break the resonance). Some people swear they can hear radio stations via their metal teeth fillings.

Now back to light. One major difference between visible light and FM radio is that the former has a far higher frequency, by about a factor of a million. Car radio circuits, or any electrical circuit for that matter, cannot resonate at frequencies this high, partly because it is difficult to build parts that small. They can resonate in the microwave range (e.g., your cell phone), but this is still about one thousand times too slow. There have been some fascinating attempts to build nano-sized electrical circuits that oscillate at optical frequencies (e.g., Engheta et al., 2005), but it is quite the challenge.

Part of the problem is that the wavelengths of visible light are also about a million times smaller than those for FM radio, which means you need small antennas. Fortunately for us, nature has provided antennas with the right size and properties. They are called atoms and molecules (figure 4.3). While visible light waves oscillate far too quickly to create any bulk movement of electrons, they have just the right frequency to vibrate electrons within their own orbits. This is hard to picture, since electrons don't orbit the nucleus in quite the way that the earth orbits the sun. Instead it is best to think of the electron as a cloud surrounding the nucleus. As the electric field from a light wave passes an atom, the negatively charged electron cloud gets pulled one way and the positively charged nucleus gets pulled the other way. Then, very quickly, both the cloud and the nucleus get pulled in the opposite direction. This repeats over and over, and the relative positions of the electron cloud and nucleus (together called a "dipole") oscillate at about a hundred trillion times a second.

In all cases, the oscillating dipole emits an electromagnetic wave. We call this scattering. However, just like clock pendulums, water in bathtubs, and Dodge Darts, atoms and molecules have resonant frequencies. At these frequencies, the incident visible light wave does a lot of work on the dipole and thus loses energy. In other words, the light gets absorbed. It is important to note that, although the incident light is being absorbed, it is driving the dipole through large oscillations, which is in turn emitting a wave. This is known as "resonant scattering." It won't be discussed further, but it is important to realize that scattering and absorption are not opposite sides of the same coin. They can and do occur at the same time.

So that, in a nutshell, is the classical view of scattering and absorption of light by matter. I personally find it intuitive and satisfying, though it has its limits. One big limitation is that it is probably not fair to assume that atomic-scale objects behave like tiny clocks and pendulums. The other is that the

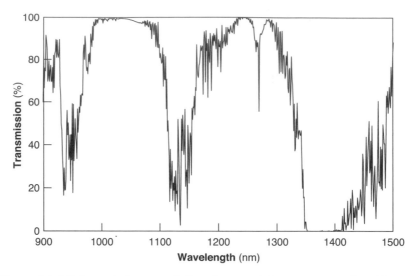

Figure 4.4: High-resolution near-infrared transmission spectrum of earth's atmosphere. Each spike-like drop is due to the absorption of light by a molecular component of the atmosphere (usually water vapor in this wavelength range). Based on the standard reference spectra measured by the American Society for Testing and Materials.

classical explanation cannot predict at what wavelengths/frequencies these resonances occur. No atom, even one as simple as hydrogen with its one proton and electron, has just one resonant frequency. Like my dad's car, they have a great many. This means that they have multiple narrow spectral regions of absorption called absorption lines.Molecules and real-life bulk matter have a veritable forest of them (figure 4.4).

Quantum mechanics does not impose classical mechanisms on atomic-scale bodies. More importantly, it can, at least in principle, predict where the resonances are. I say "in principle," because the actual calculations for anything beyond isolated atoms are quite challenging. Also, as we discuss in chapter 10, the theory has some profoundly odd characteristics. However, the accuracy and consistency of the results in general outweigh the non-intuitive implications of the theory. It is a Faustian bargain, a bit like being given the ability to predict lottery numbers at the expense of accepting that Elvis may be trapped inside your pinky.

In quantum mechanics, atoms and molecules are said to have a complex set of possible discrete energy levels, usually depicted as a set of ladders with unevenly spaced rungs (figure 4.5). To make things simple, let's assume we are dealing with an isolated hydrogen atom and that it initially is in its lowest energy state, called the "ground state." A photon flies by and strikes the atom. If the energy of the photon is equal to the difference between the energy of the ground state and the energy in a higher possible level, then the photon will be absorbed and the electron will move to an orbital with a higher energy. A short time later, the energy level of the electron will drop back down to the ground state. If it does it in one jump, it will emit a photon with the same energy as the incident photon. This is resonant scattering. However it could also get back down to the ground state in smaller jumps, like a slinky going down stairs. After the last jump, the atom will emit a photon with an energy equal to the energy difference of that jump, which will be less than the energy of the original photon. So the emitted photons will each have a longer wavelength than the absorbed one. This is fluorescence, which we will discuss in chapter 7. The atom can also travel back to the ground state via many tiny jumps and emit no visible light at all. This is what we call absorption. What path the electron takes on its trip back down to ground depends on the atom and its environment (gas, solid, molecular, or atomic, etc.), with the devil very much being in the details.

Like so much of optics, the classical and quantum-mechanical explanations of absorption each have their pros and cons. Quantum mechanics agrees exceedingly well with experimental data, but offers little intuitive sense of what is going on. The classical theory is not as precise and fails to explain atomic processes, but provides an explanation that is easily understood and meshes well with the rest of physics. Both are well-developed metaphors for processes we will never actually see. For biologists, both metaphors are equally worthy, though for most physicists and chemists, quantum mechanics is the clear winner. You can go with one or the other if you like, but I find it easier to jump back and forth. They each have their own explanatory power. While they seem entirely separate, you will find that, like chocolate and peanut butter, they go together better than you might expect.

Regardless of how you think about absorption, it is important to realize that not all energy levels involve jumps between electron orbitals. Molecules also vibrate and rotate, and these motions, like any other oscillation also have resonances with discrete energies. The energy levels for vibrations and rotations are closer to each other than those between different electron orbitals,

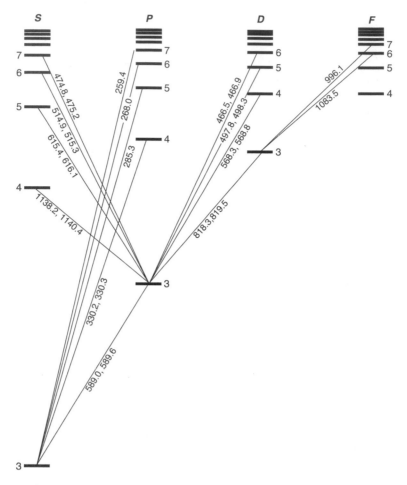

Figure 4.5: Energy diagram of atomic sodium, showing the allowed jumps and the wavelength of light that is emitted as the energy level drops. S, P, D, and F refer to atomic orbitals. There are two similar numbers for each drop because there is a fine structure that isn't shown in this figure. Courtesy of John Boffard.

so they absorb light with lower energies and longer wavelengths. In most organic molecules, the differences between vibrational and rotational energy levels correspond to infrared radiation. In these same molecules, the differences between the energy levels of different electron orbitals correspond to ultraviolet radiation. There is a window of wavelengths of light that are too energetic to match vibrations and rotations and too weak to kick electrons into higher orbits. In this range (from about 350 nm to 750 nm), there is little absorption by water and most organic molecules. This relatively tiny window between these two highly absorbing regions is what we call visible light. It is a complete accident that we can see at all.

Also it is critically important to realize that the absorption characteristics of a substance depend greatly on how it is assembled. The passenger seat in my dad's car vibrated wildly when the car drove at 55 mph, but the same seat removed from the car and pushed down the road would vibrate at a different speed. Similarly, the absorption spectra of atomic hydrogen are different from those of molecular hydrogen and quite different from those of hydrogen atoms bound up in organic molecules. A wonderful example of the effect of environment and assembly is the difference between diamond and coal. Both are made of nearly pure carbon, but the former is transparent while the latter is strongly absorbent. The difference lies in how the carbon atoms are bonded together.

Unfortunately, predicting the absorption characteristics of a complex molecule based on its structure is difficult. Like love, molecular absorption is simple in principle, but difficult in practice. In fact, both the classical and quantum explanations of absorption are a bit like the Monty Python skit where the entire lesson on how to play the flute is "blow into the big hole and run your fingers up and down the other holes."

Absorptivity, Absorptance, Absorption, Absorbance, Absorbilicious

Before we get into the day-to-day implications of absorption, we need to deal with some terminology. While the various terms related to light absorption don't have the baroque insanity of photometric units, they can still be confusing. So, taking them in order, let's start with absorptivity.

We introduced absorptivity in chapter 3 when discussing black body radiation. Absorptivity is a dimensionless number between zero and one that tells you what fraction of the incident radiation is absorbed. It is a property of a whole object, depending not only on the material, but on the geometry. Remember that, in chapter 3, we discussed how a cavity with shiny walls could still have a high absorptivity. Absorptivity is also primarily a term used by people who work on thermodynamics. So, unless you work on the heat budgets of animals, you are unlikely to use "absorptivity" often.

Absorptance and absorption are synonymous with absorptivity, but are more commonly used by biologists. Absorptance is meant to complement "transmittance," but has always had an odd sound to me. I, and many people who study the absorption of light by photoreceptors, water, and other common objects, prefer to use "absorption." Like absorptivity, absorption depends on geometry. A thicker slab of black glass has a higher absorption than a thinner slab of the same glass. Something can also have high absorption without necessarily being made of a typically absorbing substance. For example, a stack of razor blades viewed from the side looks amazingly black, even though the blades are shiny, because the light bounces back and forth between the blades, eventually getting absorbed.

Now for the term that truly is different—absorbance. Absorbance is given by:

$$A = -\ln\left(f_T\right),$$ 4.1

where f_T is the fraction of light that is transmitted and "ln" stands for natural logarithm (logarithm with a base of e). At first, this measurement makes no sense. Why bother to take the logarithm and why look at the fraction of light that is transmitted? Isn't this supposed to be about absorption? There is actually a good reason for this and it has to do with the way light is absorbed.

Imagine you have a slab of dark glass one centimeter thick. You shine a parallel beam of light through it and find that only 50% of the light makes it through. So how much makes it through two centimeters of this glass? While you might at first suspect that the remaining 50% of the original light gets absorbed (leaving you with nothing), in fact it is 50% of the 50% that made it through the first centimeter, so 25% of the light gets through. In other words, the light gets absorbed exponentially. This is not a property that is unique to light, but is found in many situations where events happen randomly, for example the half-lives of radioactive materials. Just like a radioactive nucleus has a constant probability of decaying at any moment, a photon traveling through a homogeneous material has a constant probability of striking a molecule and being absorbed. As you remember from precalculus, anything that happens exponentially looks linear in logarithmic space. In the case of light absorption, if you line up two objects and shine a beam of light through them, the total absorbance is just the sum of the absorbances of the two objects. Also, since what truly matters is how many absorbing

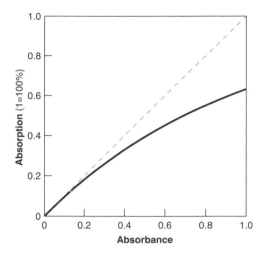

Figure 4.6: The corresponding absorption for a given absorbance (the dotted line shows absorbance equaling absorption). The two are equal for small values and then diverge.

molecules the light interacts with, doubling the concentration of a solution doubles its absorbance. This is convenient in the same way slide rules were convenient for multiplying numbers (because they added the logarithms of two numbers).

Unfortunately, absorbance gets confused with absorption (and the synonymous absorptance and absorptivity). In fact, this may be one of the most common misidentifications in optics. Oddly enough though, in many cases it doesn't matter, because, when light absorption is small, absorption and absorbance are numerically equal, even though the former is the fraction of the light absorbed and the latter is the negative logarithm of the fraction of light transmitted. This happens because, when a number is small, the logarithm of one plus that number equals just that number, which is to say:

$$\ln(1+x) \cong x \text{ when } x \text{ is small.} \qquad 4.2$$

So, if absorption is small, then:

$$A = -\ln(f_T) = -\ln(1 - f_A) \cong f_A, \qquad 4.3$$

where f_A is the fraction of light that is absorbed, which is absorption/absorptance/absorptivity. While this approximation works well if absorption is small, it quickly breaks down at higher values (figure 4.6). For example, absorbance is only 5% higher than absorption when the latter is 10%, but 40%

higher than absorption when the latter is 50%. Even more important, absorbances add while absorptions sort of multiply. So, supposing you have a value x for one centimeter depth of water and want to figure out what it would be for one meter, you had better know whether that value is absorption or absorbance. The absorbance for that one meter would be $100x$ but the absorption would be $1-(1-x)^{100}$ (we'll explain why in a moment).

A final confusing fact about absorbance is that it comes in a second flavor, involving base 10 logarithms instead of natural logarithms. This form is called "optical density" (OD) and is given by:

$$OD = -\log_{10}\left(f_T\right) \hspace{4cm} 4.4$$

So, an OD of one means that only 10% of the incident light is transmitted, an OD of two means that only 1% is transmitted, and so on. Optical density is mostly used by analytical chemists, some molecular biologists, photographers, and the makers and users of optical filters. The natural logarithm formulation is more commonly used by vision scientists, oceanographers, and atmospheric scientists. Unfortunately, many papers, people, and spectrometers do not tell you which form of absorbance they are using, so be careful. The two forms only differ by a multiplicative constant (OD = $A\log_{10}e$ $\cong 0.43A$), but this difference can be enough to confuse things.

Going from Absorbance to Absorption: The Absorption Coefficient

Suppose you want to know how much blue-green light is absorbed after passing through ten meters of lake water. Or suppose you want to know how much more light is absorbed by a photoreceptor if it grows twice as long. You could get a blue-green laser and shoot it through all that water, and you could measure the absorption for both lengths of photoreceptor, but in a lot of cases this is inconvenient or impossible. In many situations, you can only get a measurement for one length. Luckily, you can use this measurement to get a good estimate for both longer and shorter lengths.

The process works like this. Suppose you have a slab of material of some characteristic thickness, let's say one cm. Now shine a light of a given wavelength through the material. How much of it is absorbed? Assuming that there are not any significant reflections or scattering within the slab, all the light is either transmitted or absorbed. So, the fraction of light absorbed is:

$$f_A = 1 - f_T.$$

4.5

Inverting equation 4.1 gives you:

$$f_T = e^{-a},$$

4.6

where a is the absorbance of the one cm slab. So,

$$f_A = 1 - e^{-a}.$$

4.7

Now what if the slab is not one cm thick? Suppose it has a thickness of d cm. We know that absorbances add, so the absorbance of the new slab is ad. This means that:

$$f_A = 1 - e^{-ad}.$$

4.8

This absorbance, usually defined over one centimeter, is called the "absorption coefficient," and it and equation 4.8 are incredibly useful. If you know the absorption coefficient you can measure how much light is absorbed over any distance. For example, you can measure the absorption coefficient of lake water in a cuvette in a lab-based spectrometer and use it to determine how much light is absorbed over far longer distances (assuming the water is optically homogeneous).

You can also use it to determine how changing photoreceptor length affects light absorption. Using this example, you can see that light absorption is not entirely intuitive. Of course, all other things being equal, longer photoreceptors absorb more light. Indeed, it has long been known that many deep-sea species have exceptionally long photoreceptors. However, because of the form of equation 4.8, making your photoreceptor longer and longer eventually leads to diminishing returns. For example, the rod outer segments (ROS; the portions of the photoreceptors that contain the visual pigment) in the retina of the deep-sea hatchetfish *Argyropelecus* sp. have an absorption coefficient of 0.064 per micron for the blue-green light that predominates at depth. If its rods were 10 μm long, they would absorb 47% of the incident light. It's dark down there, so let's assume that the fish would like to absorb at least 99% of the blue-green light that enters its eye. You might guess that doubling the length of the ROS would do it, but this only gets you to 72%. Quadrupling the length to 40 μm only gets you to 92%. To get above 99%, the ROS must be at least 75 μm long, nearly eight times as long as an ROS that already absorbs nearly 50% of the light. So you can see that getting the last few photons involves considerably more effort than getting the first few.

This effect is not limited to photoreceptors, but is a general characteristic of light absorption. Absorbing the last few photons always takes much longer than you might expect. This is why, even though only 10% of daylight remains at a depth of 70 meters in clear ocean water, you (as a dark-adapted human) can still see some light down to about 850 meters depth.

An important thing about absorption coefficients is that—unlike all the other measures of absorption we have discussed—they have units, which are usually based on the characteristic length over which the absorbance was measured. In many cases it is cm^{-1}, but not always. For example, in oceanography it is usually m^{-1}, because light absorption by water over one cm is usually too small to be useful. In vision research, it is often given in μm^{-1}, because photoreceptor lengths are on the order of microns. An extremely common mistake is to mix units, for example using an absorption coefficient in cm^{-1} and a distance in meters. A good rule of thumb is that anything in an exponent has to be unitless to make physical sense. So make sure your units cancel out.

Now that we have looked at the physics of light absorption and squared away the units, let's look at what actually happens with absorption in nature, beginning with color.

ABSORPTION AND SCATTERING

Before I realized I was terrible at it, I was fascinated with the game Go. I went so far as to buy a set of Yunzi stones for the game. These are made in China using a secret process and come in the two traditional colors of black and white. The black stones have an especially rich darkness to them. One day, while I was losing to a bored guy who was also playing eight other people at the same time (and barely looking in my direction), I picked up one of the black stones and held it against the light. It was brilliant green. I put it back on the table and it was black again. I did this a few more times, conceded the game, and went home.

The black/green Yunzi stones display a central fact about pigments and absorption that is overlooked by many biologists—pigments never make light, they only take it away. Sounds obvious, but how many times have you heard someone say that they painted a white room another color to brighten it up? Or—closer to home for many biologists—that a plant or animal has evolved a pigment to increase its reflectance of a certain color? This is seen particularly often in papers about UV colors, where the authors measure high reflectance in some portion of the UV and then posit that the organism has added something to increase its reflectance in this region. More indirectly, but equally inaccurately, people name pigments by the wavelengths they don't absorb, rather than by the wavelengths they do absorb.

I think the root of this problem lies in the human tendency to focus on what is present, rather than what is absent. We characterize chlorophyll by its lack of absorption of green light, not by its strong absorption of red and blue light. In essence, when it comes to pigments, we look at the world upside down, which often gives us a poor understanding of what is truly going on.

Without scattering, all pigmented substances are black when viewed via the light they reflect. If you bought some paint that was pure pigment and tried to paint over a black wall, the wall would still look black. Of course, if you shine light through the pigment, you will see that it is colored, but the vast majority of biologically relevant objects are viewed via reflected light, not transmitted light. Even most transparent species are primarily detected by reflected light, not by light transmitted through their bodies. The most common counterexamples I know of are leaves viewed from below via transmitted sunlight.

Barring backlit leaves and a few other things, you are mostly left in a world where pigment alone has little effect. The Yunzi stones I bought are partially made of jade and highly pigmented. However, because they scatter little light and absorb so strongly that you cannot even see the white of a table below them, they look black.

Anyone who has taken oil painting classes also has experience with this, because many oil paints look nearly black when they come out of the tube. It's not until they are mixed with white paint that they look like the color on the label. White paint (and white in general) is more special than it looks. Rather than simply being paint without pigment, it is actually a mixture of a

transparent latex or oil base (that holds it to the wall) and powdered titanium dioxide. Titanium dioxide has a tremendously high refractive index in
the visible range (~2.5), about the same as diamond. So a powder made from
it scatters a lot of light.

We will talk more about this in chapter 5, but the important point for
now is that white paint scatters light so well that much of the incident light
comes right back out of the painted wall, giving it a high reflectance. Add it
to pure pigment and you will actually be able to see the color. Add too much,
though, and the color looks washed out. As with much of life, there is a balance. You need enough white paint that a significant amount of light is scattered back to your eye (otherwise the wall will look black), but not so much
that light is scattered back out of the wall before it has had time to be significantly modified by the absorbing pigment. The easiest way to see this for
yourself is to get a tube of oil paint (some dark color like dark green is best)
and mix it with increasing amounts of white paint. You will see it go from
black, to increasingly brighter green, to pale green, and then to off-white.

This is also true in biological coloration. Unless an organism is viewed via
transmitted light, all the pigments in the world won't make it colorful unless
its tissues also scatter light. Luckily, large organisms have a simple way of
doing this—reflection by their underlying tissues. As we will discuss further
in chapter 6, reflection is just a form of scattering, and the connective tissues
of large, complex organisms (roughly goldfish grade and above) are highly
reflective. This is why albino humans and unpigmented hair are white and
not transparent. Organisms of this size can simply place pigment on top of
these reflective layers. The reflected light will be modified, and the animal
will be colored.

But what about smaller organisms that would be transparent without
pigment? Or tissues of larger organisms that are too thin to reflect much
light on their own—for example bird feathers and butterfly wings? These
organisms and tissues have two choices. First, they can use structural colors,
where the color comes from constructive interference from repetitive structures. This will be discussed more in chapter 6. For now, it is important to
realize that this is a fairly common solution for small things that need to still
be highly colorful. A wonderful example of this is found in the pontellid
copepod *Sapphirina*. While the females of this genus are fairly nondescript,
the males are intensely blue. They achieve this coloration via stacked plates of
high-index guanine crystals, the same substance used in fish scales. Given

their small size, it would be nearly impossible for them to achieve this saturated color with pigments. Indeed, other pontellids that use pigments to become blue (they are a surprisingly blue group) are far less striking.

The other solution is to combine pigment with structures that are efficient at scattering light. Examples of this include the iridiphores in many species of cephalopod and the ultrastructure of bird feathers. In the former case, the chromatophores of squid and octopi are often underlaid with a complex network of structural reflectors. Without this underlying layer, the color made by the chromatophores would be far less distinct, since cephalopod muscle tissue is often quite transparent. In the latter case, bird feathers have a sponge-like ultrastructure, where keratin alternates with air. This, of course, makes the feathers lighter, but it also scatters a great deal of light, which is why even thin feathers are often white rather than transparent. Combining this ultrastructure with pigment can make even small feathers colorful.

Interestingly, the ultrastructure of both cephalopod iridophores and bird feathers can do more than just provide indiscriminate scattering. Some cephalopod iridophores are under neural control via acetylcholine and can modify their reflectance, adding a second, dynamic color component to the animal's already impressive repertoire (Uzumi et al., 2010). No bird feathers are known to dynamically change their ultrastructure in this way, but many species have organized this spongelike network of keratin and air into structural colors, creating iridescence and highly saturated hues (e.g., Prum et al. 1998, 1999).

We will continue the story of structural coloration in chapter 6. For now though, the central message is that, with the exception of rare backlighting situations, color in nature requires both absorption and scattering in the right balance. Too much of one or the other and you only have black or white.

ABSORPTION AND COLOR

When I was a kid, I dyed Easter eggs. After a short hiatus, I started back up again with my own kid. If you have ever done this, you may have noticed a funny thing when you looked at the dyes in the bowls (before they go on the egg). The blue dye looks blue and the green dye looks green, but the yellow dye looks red. This funny fact actually matters biologically. It also illustrates why you need to think about pigments in terms of what they absorb, not what color you see. So I am going to explain it in some detail.

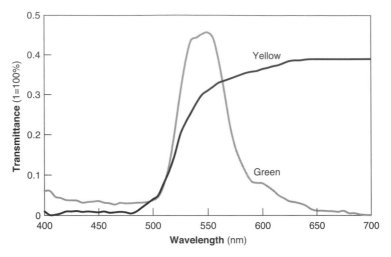

Figure 4.7: Transmission spectra of hypothetical yellow and green egg dyes.

Figure 4.7 shows you the transmission spectra of fairly typical green and yellow dyes (let's assume that the light has traveled through 1 cm of each dye). As you might expect, the green dye has a peak in the green portion of the spectrum and has low transmission everywhere else. The yellow dye is different, though. Rather than having a peak in the yellow, the transmission is low until about 500 nm, rises quickly from 500 nm to 580 nm and then stays at a high level for the rest of the spectrum. This sort of spectrum has a couple of different names. Some call it a step function (for obvious reasons), others call it a high-pass filter, because it only transmits light above a certain wavelength. Both are a bit of a misnomer, though, because if you looked just a little into the UV, you would see the reflectance go back up. Whatever you call it, though, nearly all yellows, oranges, and reds in nature have this shape. For unknown reasons, you almost never see a yellow, orange, or red color that has a simple "hump" spectrum, like you see with green. Instead you see a step function, with the location of the step determining which hue you see.

Nearly always, this step function results from the absorption properties of a large family of molecules known as "carotenoids." There are about six hundred of these long, skinny molecules, made by plants and some fungi. With the exception of one species of aphid (that apparently acquired the genes for synthesis from fungi), animals cannot produce carotenoids, but must eat

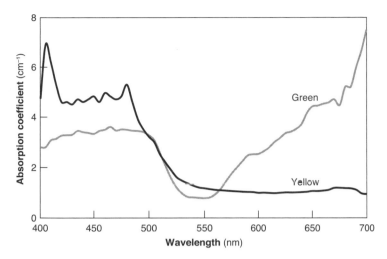

Figure 4.8: Absorption coefficients of the two egg dyes shown in figure 4.7.

them (Moran and Jarvik, 2010). These molecules play a host of roles (including being the precursor of the retinal chromophore in our visual pigments), but our main interest here is that they provide many organisms with their yellow, orange, and red coloration.

Going back to the egg dyes, let's look at spectra of their absorption instead of their reflectance. Better yet, let's look at their absorption coefficient over a characteristic length of one cm (figure 4.8). Looking at it this way, you see that what makes the green dye green is that it absorbs nearly all the blue and red light, leaving only the green light to be reflected. Chlorophyll does the same thing, absorbing light strongly at both ends of the spectrum, but not in the middle. You can also see that the yellow dye strongly absorbs short-wavelength light, but absorbs long-wavelength light less. Both dyes actually absorb a fair bit of light from all wavelengths, which means that they are fairly concentrated.

Using equation 4.8, let's see what happens to the color we see as we pour the dyes into bowls with white bottoms. The white bottoms act as the reflectors in this case and so the light that reaches our eye is what is left after being transmitted through the depth of the dyes twice (on the way in and on the way out). Because we are ignoring any scattering (aside from reflection from the bottom), the transmitted and absorbed light must add up to all the original light, so:

$$f_T = 1 - f_A = 1 - \left(1 - e^{-ad}\right) = e^{-ad}.$$

4.9

In this case d is twice the depth of the dye in the bowl. Figure 4.9 shows what happens to both the green and yellow dyes as we increase the amount of dye in the bowl. In the case of the green dye, the peak goes down, so the dye gets darker. However, the peak remains in the same place, so the color does not change. It does become more saturated (or more pure, as some people call it), but the hue remains the same. The yellow dye becomes darker as well, but—more importantly—the location of the step (where it goes from low to high reflectance) shifts to longer wavelengths. Because light is exponentially absorbed, regions of the spectrum that already have low reflectance are disproportionally affected by any increase in path length. For example, consider a 1-cm-thick sheet of glass that transmits 80% of 600 nm light and 40% of 450 nm light. If you made the glass twice as thick, it would now pass 64% (= 80%×80%) of the 600 nm light and 16% (= 40% × 40%) of the 450 nm light. So the 600 nm light only got a bit dimmer, but the 450 nm light was cut significantly. In the case of the yellow dye, this means that the low-reflectance end of the step function is more strongly affected by the increasing path length, which pushes the location of the step to longer wavelengths.

So what you see, as you add "yellow" dye to the bowl, is that it goes from clear to light, unsaturated yellow to more saturated, darker yellow to orange to red to dark saturated red. As path length increases, the reflectance at all wavelengths drops, the saturation of the color increases, and the hue shifts from yellow to red. The same thing also happens if you increase the concentration of the dye. What truly matters is not the path length, but the number of dye molecules that the light intercepts.

All of this matters biologically, because carotenoid pigmentation is at the center of many discussions about honest signaling, the idea being that, because most carotenoids are derived from diet, colors based on carotenoids are an honest indicator of an animal's success in foraging. Many studies linking carotenoid coloration with mate preference have been done regarding a number of species, with much of the earlier work done on guppies (e.g., Kodric-Brown, 1989). A problem with some of these studies is that they attempt to look at hue, saturation, and brightness separately, when—as we have just seen—they are connected. An increase in the concentration of the pigment changes all three in an interdependent way.

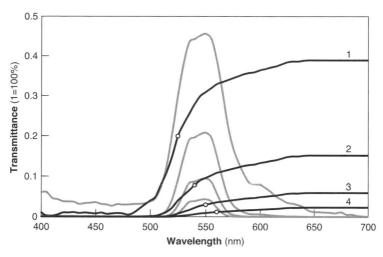

Figure 4.9: The transmittance of increasing amounts the yellow and green egg dyes from figure 4.7 (solid and dotted lines, respectively). The numbers show the path length of the light through the dye in the bowl (which is double the depth of the dye) in cm. The circles on the yellow spectra show the halfway point of the "step" and how it moves to longer wavelengths with increasing depth of the dye.

A similar thing can happen even in pigments that don't have a step function. For example, I have a pot of petunias on my back deck. When the petunias are open, they are a light purple, but when they are closed, they are blue. These flowers have a large reflectance peak in the blue portion of the spectrum and a smaller one in the red portion. In low concentrations or over the short path length of an open flower, only the middle wavelengths are strongly absorbed and the flowers look purple. However, if the concentration of pigment or path length increases (as happens when the petals are closed) the absorption increases, which has a disproportionate effect on portions of the spectrum that already have low reflectance (in this case the red). So all that is left is reflectance at low wavelengths, and the closed flowers look blue. In general, one must be careful when looking at color changes. Just because something changes color or has a color with two reflectance peaks, does not mean that it has two pigments. It is possible that everything is due to changing concentrations of one pigment.

Blue pigments can also undergo hue shifts with increasing concentration, though the effect usually isn't as strong. The reflectance spectra of most blue pigments are not step functions like carotenoids, but neither are they bump spectra like chlorophyll. Instead they are more like steeply sloping ramps. As concentration of the pigment increases, the low reflectance portions of the ramp are disproportionally affected and the hue can shift from blue to violet. This effect is less dramatic than that of carotenoids because human-perceived hues don't shift as rapidly in that portion of the spectrum. Also we tend to see true violet (high reflectance only in the shortest of human-visible wavelengths) as quite dark, which makes color shifts harder to see. However, if you look carefully, you will notice that Easter egg dye that looks blue on the egg is more violet in the liquid form (due to longer path length). The same is true for tie-dyes, blueberries, and also for the coloration of some of the intensely blue neustonic animals such as Portuguese man-o-war (*Physalia* sp.).

SEEING THINGS UPSIDE DOWN

As I mentioned, our human tendency to concentrate on what is present rather than what is absent means that we look at pigments upside-down, focusing on what they don't absorb, rather than on what they do absorb. This can get us into trouble in many ways.

First, let's go back to kindergarten. Maybe now they teach reading and math at age five, but when I was a kid, the first thing we learned was that blue and yellow add to make green. However, if you add the reflectance spectra of typical blue and yellow paints, you don't get a reflectance spectrum that looks green at all. In fact, if mixing paints worked like this, yellow and blue would make purple! To determine what happens when you mix paints (or any pigments), you need to multiply the reflectance spectra. This is assuming that our kindergarten teacher gave us some nice white paper (instead of the flimsy, brown stuff we usually got), so that the reflectance is closely related to the fraction of light that the paints transmit.

If you do this for yellow and blue paint, you find that the product is low at short wavelengths (because yellow paint transmits little short-wavelength light) and low at long wavelengths (because blue paint transmits little long-wavelength light). The only part of the spectrum where the product is above zero is in the middle, where both paints transmit some light, so the mixture looks green (figure 4.10). This only works if you have some overlap in the two reflectance spectra. If your blue and yellow paint were highly saturated

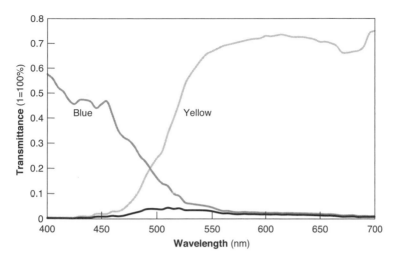

Figure 4.10: Mixing yellow and blue paints (dotted lines) to make green (solid line).

and didn't have any overlap in the middle of the spectrum, their mixture would be black. This actually happens with some high-quality oil paints, proving that sometimes even quality can be too much of a good thing.

The other thing to realize is that, at every wavelength, the reflectance of the mixture will be lower than that of either original color. Pigments can only take light away. I spent a lot of time painting from kindergarten through graduate school and can tell you that, while blue and yellow do make green, they make a dark and muddy green. The crueler art teachers would make you paint with only red, blue, and yellow, saying that we could mix any color we wanted. True, if we wanted all the colors to look like mud. Most of us snuck tubes of real green, purple, and orange into class.

Moving from art to animal coloration, let's suppose you want to put green spots on a red frog. Suppose also that the red in the frog comes from a carotenoid and is saturated enough that it transmits no light with a wavelength below 600 nm. If you mix green pigment in with the red, the product will still be zero at wavelengths below 600 nm, so the best you'll end up doing is making the red darker (because the green pigment will absorb some red light). We also can't just put the green pigment on top of the red pigment, because the light hitting the green pigment won't get back to our eyes until it's passed through the red pigment, hit the underlying reflective connective tissue, and gone through the red layer again. So you'll still only end up with a darker red spot. There are only two solutions. You could get rid of the red pigment wherever you want to add green pigment. This obviously works, but adds a step that complicates the evolution of color change. Or, you add something to the green pigment that scatters light efficiently and then layer that mixture over the red pigment. This way, a significant amount of the incident light is scattered back out of the frog before it has a chance to interact with the underlying red layer. This scattering could be achieved by putting a reflective layer between the green and red layers, adding high refractive particles to the green layer, or making the green a structural color. In chapter 6, we'll look at examples of animals that do these things.

A second example of the importance of looking at animal coloration from the perspective of absorption can be found in the deep sea. If you look at photos of deep-sea species, you notice a couple of unusual things. First, that, while many of the species (mostly the vertebrates) are black, most (especially the invertebrates) are red. You will also notice that, of the red taxa, the pelagic species have far more saturated colors than the benthic ones. The pelagic ones are generally dark and pure red, while the benthic species ranges from orange-red to salmon-colored to a tan-yellow.

As you may know, clear water absorbs blue-green light the least. The difference in absorption is fairly small over short wavelengths, which is why a glass of water has no color, but, because light is attenuated exponentially, these differences become enormous after you've gone through a few hundred meters of water. So, by the time you get to depths of 500 m or so, the remaining light is a blue-green with almost laserlike purity. The local animals have adapted to this situation and have visual pigments that are most sensitive to blue-green light. For these two reasons, all that truly matters for animal coloration is what happens in the blue-green portion of the spectrum.

It turns out that, if you look at the benthic animals in their habitat using just blue-green light (by either putting blue-green lights on your submarine or looking at only the blue channel of your photographs), you see that they match their substrates quite well. Those that are orange-red have relatively low reflectance in the blue-green and thus best match darker substrates, like dark mud. Those that are yellow have relatively high reflectances in the blue-green and best match lighter substrates like sand and coral rubble (figure 4.11). In other words, the colors we actually see are completely unimportant to the animal. The pigment is just there to modulate the animal's reflectance

Figure 4.11: *In situ* image of the deep-sea crab *Munidopsis* sp. and the arm of an unidentified ophiuroid under red (left) and the more natural blue (right) illumination. Both animals are more obvious under red light.

in the blue-green portion of the spectrum. If this was done via a more spectrally neutral pigment like melanin, the match would be more obvious to us, even in full-color photographs, but the red fools us. Even if we know that it's only the reflectance in the blue-green that matters, it is nearly impossible to tell how well the animal matches the background in that wavelength range when faced with all that red.

So that explains what's happening with the benthic animals. What about the pelagic species? Why are they such a dark and saturated red? In this case, there is so much red pigment that the reflectance in the blue-green is essentially zero. You might think that this is to match the dark background, but the reflectance is so low that, given the lighting conditions at depth, the animals are actually darker than the background water and stand out like silhouettes, much like trees against a night sky. So far, the only explanation that makes any sense is that the deep-red pelagic animals are hiding from searchlights. As mentioned in chapter 3, many deep-sea predators have photophores directly under their eyes that shine forward and potentially illuminate prey (figure 4.12). The light from these photophores is blue-green. Since the water itself scatters little light back to the viewer, any reflectance from the animal will stand out and be worth investigating. Note that this is different from the benthic case where the substrate also reflects a fair bit of light. So the best solution for a pelagic prey animal is to have zero reflectance in the blue-green portion of the spectrum (figure 4.13). Some species do this with melanin and are black, but many do it with carotenoids and so are deep red.

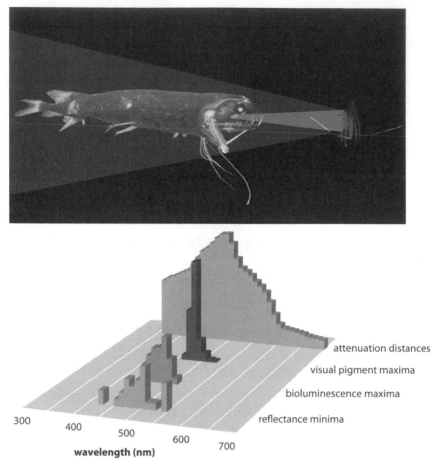

Figure 4.12: (Left) The dragonfish *Aristostomias* sp. illuminating a ctenophore with its ocular photophore (fish image courtesy of Edith Widder). (Right) Histograms showing the wavelengths at which the reflectance of deep-sea pelagic animals is minimal combined with histograms of bioluminescence and visual pigment maxima for deep-sea animals. The rearmost plot is not a histogram but a graph of the attenuation distance (distance over which radiance drops to $1/e$ or 37% of its original value) of the water as a function of wavelength.

Interestingly, even the deep-dwelling relatives of transparent taxa often have a layer of red pigment over their still-transparent bodies. This occurs because transparent objects can be easily detected by flashlights in the dark. If you don't believe this, go out tonight and shine a flashlight at a darkened

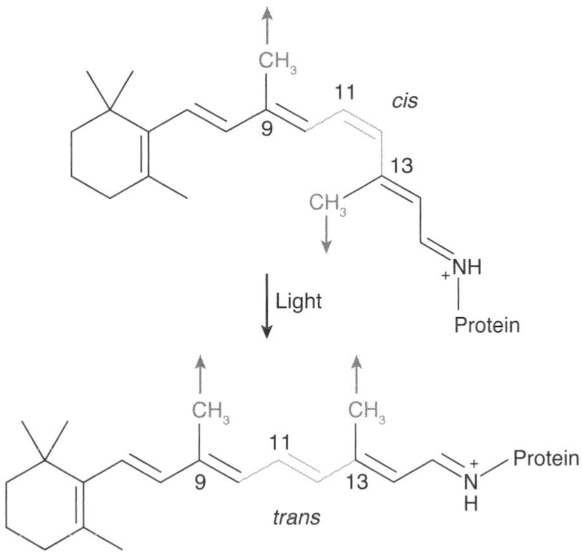

Figure 4.13: Change of retinal chromophore shape after absorbing a photon. After Liu et al., 2007.

window. You see a strong reflection of your light because the window has a different refractive index than air and reflects light. The difference in refractive index between animal tissue and water is smaller, but still enough to reflect light. So transparent animals viewed by searchlights are paradoxically more visible than those that absorb light.

The obvious next step for any flashlight-bearing predator is to have a red flashlight. You still wouldn't see the black melanin-coated animals, but the red animals would light up like Christmas trees. Surprisingly, only three deep-sea genera are known to do this: *Pachystomias*, *Aristosotmias*, and *Malacosteus*—all members of the dragonfish family (Stomiidae). The nine species of this group use dense filters on their flashlights to convert green-yellow bioluminescence to red (making it much dimmer in the process). It's not just any red either, but one that is shifted nearly into the infrared. For example, the peak emission of the bioluminescence of *Malacosteus niger* is 705 nm. It is not known for sure if this red bioluminescence is used to detect red prey or for communication (or both). However it is a successful solution to finding red prey at depth.

But how do these animals see their own light? This brings us to one of the most important instances of absorption in biology—vision.

VISUAL PIGMENTS

As I mentioned at the beginning of this chapter, without absorption you can't see. Visual pigments are also a wonderful example of how the structure of a molecule affects how it absorbs light. While I cannot possibly do justice to the vast and fascinating literature on the absorption of light by photoreceptors, I would like to at least get across a few of the fundamental principles.

With the exception of phytochromes and cryptochromes, all photoreception is based more or less on the same system, which consists of a small molecule bound to a protein. The small molecule, called the "chromophore," is a carotenoid derivative of vitamin A. The most commonly used one in both vertebrates and invertebrates is an aldehyde of vitamin A called "retinal" (A_1). Next most common are the closely related 3,4-didehydroretinal, found in some fish, amphibians, reptiles, and crayfish (A_2) and 3-hydroxyretinal (A_3), found in many insects, including flies and butterflies. The deep-sea squid *Watasenia scintillians*, in addition to A_1 and A_2, has a third chromophore, (4R)-4-hydroxyretinal (A_4).

These four chromophores all have slightly different absorption characteristics. In some cases multiple chromophores are found in the same animal, sometimes at the same time, sometimes under different conditions or stages of life. In the case of *W. scintillians*, the three chromophores appear to give it color vision, the only example of such among the cephalopods (Seidou et al., 1990). Since most animals use A_1, and since the basic principles of light absorption are independent of chromophore type, we will stick to A_1 for the rest of this discussion. We'll refer to it as retinal, which is easier to say than A_1 (and sounds less like a steak sauce).

Retinal comes in two flavors, 11-*cis* and all-trans. In the 11-*cis* form, the molecule is kinked at the eleventh carbon; in the all-trans form, the molecule is straight. The absorption of a photon converts the molecule back and forth between these two states, sort of like a toggle switch (figure 4.13). However, visible photons are not good at doing this. In fact, retinal is most sensitive to photons with a wavelength of 340 nm, well into the ultraviolet. This is where the second part of the system, the protein, comes into play. The chromophore is covalently bound to a large membrane protein, known as an "opsin." Opsins are members of a large, complex family of proteins known as "seven-membrane-spanning G-protein-coupled receptors," a family that also binds odorants, neurotransmitters, and molecules of the immune system. In the case of photoreception, the opsin plays two important functions. First, when the chromophore changes shape after absorbing a photon, this change forces the opsin to change shape as well. This shape change in the opsin activates a G-protein on the inside of the cell that sends the message that light has been absorbed. In the case of vertebrates, the opsin loses its chromophore after changing shape and is called bleached until a new chromophore is attached. In invertebrates, the straightened chromophore remains attached to the opsin and is rebent by photons of a different wavelength. The second thing the opsin does is modify the absorbance spectrum of the chromophore. Just like sticking a seat into my dad's Dodge Dart changed the frequency at which the seat oscillated most easily, putting a chromophore into an opsin changes which wavelengths of light it absorbs most efficiently.

From an evolutionary standpoint, this is convenient, since selection can act on the gene for the opsin and alter the spectral sensitivity of the visual system over a wide range. From a human standpoint, though, it's difficult to predict the spectral sensitivity of a given chromophore-opsin combination from the amino-acid sequence of the opsin. We know that the amino acids directly surrounding the chromophore matter the most, but not why changing one residue from a valine to a isoleucine shifts the peak of the absorbance spectrum by 45 nm, while changing it to a cysteine only shifts it by 3 nm. It's a hot but frustrating area of research, similar in complexity to predicting a protein's folded structure from its amino acid sequence.

Conveniently, while the absorbance spectra of visual pigments (i.e., opsin-chromophore combinations) vary over a wide range, with peaks ranging from 300 nm to 630 nm, they all look similar. Figure 4.14 shows a set of four retinal-based visual pigments, peaking at 450 nm, 500 nm, 550 nm, and 600 nm. A couple of things are immediately apparent. First, each curve seems to be the sum of two curves, one that peaks in the visible and one that always peaks at 340 nm. The visible and UV peaks are known as the "α-band" and "β-band," respectively. The β-band curve is nearly identical to the absorbance spectrum of the isolated retinal chromophore, while the α-band varies depending on the opsin. If you look closely you'll notice that the α-band is slightly wider for the curves that have peaks at longer wavelengths, but that otherwise they all look similar. In fact, it has long been known that if, instead of plotting the curve against wavelength, you plot it against the quotient of the peak wavelength and wavelength (i.e., λ_{max}/λ), you always get nearly the same curve.

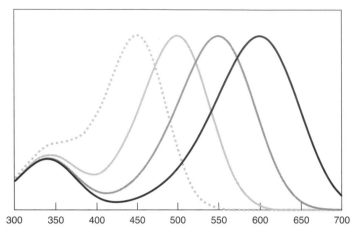

Figure 4.14: Absorbance spectra of visual pigments with alpha bands peaking at 450 nm, 500 nm, 550 nm, and 600 nm. The beta bands all peak at about 340 nm. Based on the template found in Stavenga et al., 1993; the template used in Govardovskii et al., 2000, is slightly different.

What this means is that you can accurately predict the absorbance spectrum of a visual pigment if you know two things: (1) the wavelength at which it peaks (known as "λ_{max}" and pronounced "lambda max"), and (2) the chromophore that is bound to the pigment. The latter affects the shape and peak of both the β-band and α-band. If it is a retinal-based pigment, then you only need to know the λ_{max}. There is some dickering about the relative height of the two bands, how much the peak of the β-band depends on the peak of the α-band, and other details, but the bottom line is that λ_{max} tells you just about everything you need to know (see Appendix B for detailed equations based on a commonly used template).

I have no idea why it's this simple. It seems perfectly reasonable to expect that certain opsin-chromophore combinations would have completely novel absorbance spectra, but no one has found any. It's almost like someone is doing vision scientists a favor.

These scientists may need a favor, because, while the absorbance curves are simple and essentially identical, the phototransduction process (where the original chromophore change is eventually translated into an electrical signal) is complex and varies among taxa. Again, a vast literature and several excellent books have been written on this topic (see *Invertebrate Vision* for a good start), so I won't go into it. However, many interesting optical facts remain.

First of all, as I briefly mentioned in chapter 2, photoreceptors count photons not energy. Once absorbed, a 350 nm UV photon has no more effect on a cell's membrane potential than a 700 nm red photon, even though the former has twice the energy of the latter. What the absorbance spectrum of a visual pigment tells you is not the effect a photon of a given wavelength has, but the probability that it will be absorbed. As far as photoreceptors go, all absorbed photons look alike.

Second, photoreceptor cells can actually count individual photons. Careful experiments since the 1950s have shown that a photoreceptor cell can generate an electrical response to individual photons, known as "photon bumps" (reviewed by Warrant, 2006). This sensitivity is extraordinarily good and matched only by the most sensitive of human-produced light meters—photon-counting photomultipliers. Photoreceptors not only respond to individual photons but do so fairly reliably. Although there are losses due to reflection at the cornea, absorption within the lens and other ocular media, and less than perfect transduction within the photoreceptor itself, about half the photons that strike the cornea of a dark-adapted eye create an electrical signal (reviewed by Warrant, 2006).

Given this sensitivity and reliability, you would think we would be able to see well under any light conditions, but instead our vision begins to deteriorate at levels darker than late twilight and gives out entirely at levels just below that of starlight. It's been shown that we can detect a flash of green light as dim as 400 photons/cm²/s (reviewed by Warrant, 2006), but any real vision requires light levels a couple of orders of magnitude greater than that.

There are three main reasons for this. First, and most importantly, whether you think of light as a wave or as composed of photons, its interaction with matter occurs in discrete events, much like raindrops hitting the ground. When light levels are high, this looks fairly continuous, just like rain in a toad-strangling thunderstorm. However, at lower light levels, this discrete nature, known as shot noise or photon noise, becomes apparent. How this affects your ability to see can be quantified surprisingly well because the photon interactions occur randomly and follow what are called "Poisson statistics." In random processes of this type, if the average number of events observed over a given time period is N, then the standard deviation of this average is $\sqrt{N}$. So the ratio of the number of photons absorbed to the standard deviation of this average is $N/\sqrt{N} = \sqrt{N}$.

This ratio, known as the "signal-to-noise ratio" (SNR), is an important measure of how well a given visual system can see under low light. In particular, it determines the ability to distinguish two objects of similar radiances, referred to as contrast sensitivity. Contrast sensitivity is often given by the contrast threshold, the percentage radiance difference that can just be detected. Under bright light, our contrast threshold is about 1%–2%, depending on how you measure it. As light levels decrease, this threshold rises, and is inversely proportional to the SNR. A nice rule of thumb is that the contrast threshold C_{min} is:

$$C_{min} \cong \frac{2.77}{SNR} = \frac{2.77}{\sqrt{N}} \qquad\qquad 4.10$$

where N is the number of photons absorbed per photoreceptor (Warrant, 2006). The number 2.77 is not pulled out of a hat, but comes from the relationship between the standard deviation of the mean and confidence intervals. In this case, the contrast threshold is set to be the percent difference in radiance at which you would have a 95% chance of distinguishing two areas of differing radiance.

The other two processes that limit the reliability of vision are dark noise and transducer noise. Dark noise refers to the fact that occasionally a photoreceptor will act as if a photon has been absorbed when it hasn't been. In other words, it sets off a false alarm. The dark noise in visual pigments is extremely low, about 10 events per photoreceptor per hour in insects and 360 events in vertebrates (Warrant, 2006). It increases with temperature, so the unproven suspicion is that deep-sea fish (which live at about 4°C) have especially low levels of dark noise, which in their case would be quite adaptive. A far greater problem is transducer noise. This noise is also physiological, but rather than dealing with false alarms, it refers to the fact that the electrical response to a given photon absorption is not entirely constant. This random variation is roughly equal to shot noise at low light levels, so vision is primarily limited by transducer and shot noise, with dark noise usually playing a minor role.

Together, all three processes lower the reliability of vision at low light levels. Many fascinating papers have been written on how nocturnal and deep-sea animals get around this problem. Since these are well covered in books devoted to vision (and also in chapter 9 when we talk about using spectrometers), I'll be brief. While you might think that boosting the gain in the photoreceptor would help, this boosts the noise just as much as it boosts the signal. So, all the solutions come down to the same thing—absorb more photons in the photoreceptors. This can only be done in a few ways.

Although I avoid using long equations in this book, the next one is too fundamental and illustrative to pass up. It tells you how many photons are absorbed in an individual photoreceptor. I'll first give it in a general form that shows the important factors and later in a more specific form that allows you to calculate actual photon catch. In the general form, the number of photons absorbed by a photoreceptor looking at an extended scene (i.e., not a star or other point source) is:

$$N \cong A\Omega F_{abs}\Delta t \,, \qquad\qquad 4.11$$

Let's take the terms in order. A is the area of the pupil. All other things being equal, the larger the pupil, the more light gets to each photoreceptor. This is why the major telescopes like Keck and Palomar have enormous openings; not to increase magnification but to boost sensitivity. In animals, pupil size is limited by two things: (1) a big pupil requires a big eye and thus probably a big head, and (2) a big pupil requires a big lens, which is hard to make and more likely to have aberrations. So an animal of a given size can only vary A by so much, though this does explain why nocturnal and deep-sea animals usually have big eyes for their size. Since pupils are usually circular and diameters (D) are easier to measure than radii, A is usually given by $\dfrac{\pi D^2}{4}$.

The second term, Ω, is the field of view (given as a solid angle). An individual photoreceptor will absorb more photons if it views a larger region of space. This can be done by making the photoreceptor wider or shortening the focal length of the lens in the eye. A third solution is to neurally link together a number of neighboring photoreceptors to create one super receptor (figure 4.15). This is known as "spatial summation." Because this appears to be a common tactic, it is often best to talk about how much light is absorbed by a given channel, where a channel can either be an individual photoreceptor or a set of them working as one. All these solutions lower acuity, because you have essentially reduced the number of pixels in the system. However, in

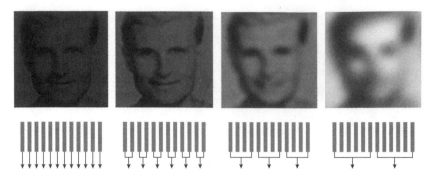

Figure 4.15: Effects of spatial summation. As more photoreceptors (gray bars) are tied together, the image gets brighter, but also coarser. Courtesy of Eric Warrant, Lund University.

many cases losing some acuity is worth the gain in sensitivity and thus reliability—no sense having millions of pixels if each one only sees noise. For example, unless the sensor chip actually gets larger, increasing the number of pixels in a digital camera lowers its ability to work in dim light. This is why sometimes buying a 6 megapixel camera is smarter than buying a 20 megapixel one. Again, since people usually prefer to think in terms of linear rather than solid angles and since fields of view are usually circular, Ω is generally given as $\dfrac{\pi R^2}{4}$, where R is the angular resolution of the channel/receptor in radians.

The third term, F_{abs}, represents the fraction of photons entering a photoreceptor that are actually absorbed by it. This depends on a photoreceptor's absorbance and length and is given by the highly useful absorption equation 4.8. In this case, it is usually written in the form:

$$F_{abs}(\lambda) = 1 - e^{-kA(\lambda)l} , \qquad\qquad 4.12$$

where l is the length of the visual pigment-containing portion of the photoreceptor, k is the absorption coefficient of the photoreceptor at the peak absorption wavelength, and $A(\lambda)$ is the absorbance of the photoreceptor, normalized to a peak of one. The variables k and $A(\lambda)$ are separated in this way for convenience. This way all the absorbance spectra have the peak value of one and you just need to keep track of the k's, which are fairly constant within a taxonomic group.

Photoreceptors are already crammed to the gills with visual pigments, so the easiest way to increase the amount of light absorbed is to increase the path length l. This can be done by increasing the length of the photoreceptor or by sending light through the cell twice. This latter trick is accomplished by putting a mirror behind the photoreceptor, so that light has two chances to get absorbed—once on the way in and once on the way out. These mirrors, known as "tapeta," are found in the eyes of many animals—shine a flashlight at your cat's face and you'll see two of them. Some animals, particularly deep-sea fish, use both tapeta and long photoreceptors to absorb nearly every photon that hits them. However, remember that, due to the exponential nature of light transmission, increasing path length eventually leads to the land of diminishing returns. At any rate, you can never get above 100% of the photons that arrive at the receptor in the first place.

The product of these first three terms is often called "optical sensitivity" (s) and written as:

$$s(\lambda) = \left(\frac{\pi}{4}\right)^2 R^2 D^2 \left(1 - e^{-kA(\lambda)l}\right) \cong 0.62 R^2 D^2 \left(1 - e^{-kA(\lambda)l}\right). \quad 4.13$$

Though it is almost never stated, it is critical to remember that optical sensitivity depends on wavelength.

The final factor affecting how many photons a receptor absorbs is Δt, the integration time. Much like a camera can control its shutter speed, a photoreceptor can control how long it samples photons before sending a signal. Increasing integration time lets in more photons, so the amount of shot noise relative to the total signal goes down. Unfortunately, it also increases the amount of dark noise and transducer noise, so it is not a perfect solution. The other problem is that increasing integration time hurts your ability to detect fast-moving objects, just like opening a camera shutter for too long can blur your image. Integration times vary over a wide range: from about 0.25 seconds for some slow-moving deep-sea animals to less than 0.01 seconds for fast-flying diurnal insects (Warrant, 2006).

Two final factors affect photon catch, but generally don't vary much. One is the transmission (τ) of the lens and other ocular media (cornea, vitreous humor, etc.). While this can be low in the UV if the animal has UV-absorbing pigments, it is generally relatively high and constant in the visible (~80%). The other factor is the quantum transduction efficiency (κ), which is the fraction of absorbed photons that actually trigger an electrical event in the cell. This has only been measured for a few species, but seems to be relatively conserved, with a value of ~50%. Putting it all together, we get that:

$$N(\lambda) \cong 0.62 R^2 D^2 \kappa\tau\left(1 - e^{-kA(\lambda)l}\right) L(\lambda)\Delta t \,, \qquad 4.14$$

where $L(\lambda)$ is the radiance of the viewed object. Since, photoreceptors aren't spectrometers and actually have broad spectral sensitivities, you usually want to know how much light was absorbed at all wavelengths. This is given by:

$$N \cong \frac{1}{4} R^2 D^2 \Delta t \int\left(1 - e^{-kA(\lambda)l}\right) L(\lambda)d\lambda \,, \qquad 4.15$$

where the integral is performed over the entire wavelength range and there is measurable light that the particular visual pigment can absorb (I also folded in the usual values of 50% and 80% for quantum efficiency and transmission to simplify things). There are further refinements one can make (based mostly on the angular sensitivity of photoreceptors and waveguide effects), but equation 4.15 gives you a good estimation of the number of photons with a minimum of fiddling. Also, as we will discuss in chapter 9, you seldom know the true light levels accurately enough to worry about smaller corrections anyway.

Equation 4.15 is extremely useful and underlies almost every calculation of photon catch you see in papers modeling visual abilities and color perception in animals. With it you can predict many things, such as contrast sensitivity and the ability to discriminate colors from one another. However, you need to be *very* careful about units. I have seen this equation used by dozens of people and can say that less than one in five get the units right the first time. The biggest mistake people make is to give R in degrees, when it should be given in radians. You can change R to degrees if you like, but remember to divide everything by 3283 (the number of square degrees in a steradian). The second mistake is to not use the same units for the diameter of the pupil as for the radiance. If the radiance is per cm^2, as it often is, then the diameter of the pupil has to be given in cm. Similarly, integration time must be given in the same units that were used for radiance, generally seconds. The last big mistake is to not use the same units for k and l in the exponent. Remember that the units in the exponent have to cancel. There is no such thing as an exponential unit. So, if the absorption coefficient is given in μm^{-1}, then the length of the photoreceptor had better be given in μm. Finally, it is useful to remember that $d\lambda$ is a real unit, so if you do your integral by adding a column of numbers in a spreadsheet program (like many of us do), make sure to multiply the sum by the wavelength interval. The bottom line for equation 4.15 is that all units need to cancel out except for photons.

As I said, there are several wonderful books on vision that go into all the subtler points of transduction, optimal spatial and temporal summation, photoreceptor waveguide theory, and many other fascinating topics, so I'll end here with one final example.

You may have noticed that I mentioned that the peaks of visual pigments range from about 300 nm to 630 nm, and you may have wondered why there aren't any visual pigments that peak farther into the red. It's a good question. There is certainly plenty of light at these longer wavelengths, at least in terrestrial environments. Also our tissues are still fairly transparent to light at these wavelengths. Stick a flashlight behind your thumb and you will see that human tissue transmits red light well.

No one knows the answer to this, but the main hypothesis is that visual pigments with peaks at wavelengths longer than 630 nm are too vulnerable to thermal radiation (Ala-Laurila et al., 2004a, 2004b). Because visual pigment absorbance spectra are so broad, a pigment with a 650 nm peak (for example) has a significant amount of sensitivity all the way up to 800 nm. The idea is that this is far enough into the infrared that thermal radiation would trigger the photoreceptor to fire. I am not entirely sure I buy this, since as you can see from chapter 3, the amount of 800 nm photons produced by objects at physiological temperatures is exceedingly small, but for the moment this is the best hypothesis we have.

So what about the deep-sea dragon fish with the red bioluminescence? The light they produce peaks at 700 nm. We can certainly see 700 nm light. In fact we can even see 800 nm light. If there is a two-photon microscope near you, ask the technician to shine the 800 nm laser onto a sheet of paper. You will see a nice red spot. However, we can only see long wavelength light if it is extremely bright. The light from the red photophores of these dragonfish is dim, so how do they see it? Some, like *Aristostomias*, have visual pigments that peak at long wavelengths. However, others like *Malacosteus* have a pigment that peaks only at 542 nm, which isn't long enough to see the 705 nm light it produces.

Amazingly enough, it looks as though *Malacosteus* gets help from chlorophyll (Douglas et al., 1998). As I mentioned before, chlorophyll is green because it strongly absorbs both blue and red light. It has a particularly prominent absorption peak at about 670 nm, close to the peak of the bioluminescence of *Malacosteus*. Based on this and the fact that chlorophyll has been selected for its ability to transfer energy to other reaction centers, Douglas et al. (1998) looked for and found a magnesium-free chlorophyll derivative in the photoreceptors of the animal. They also found that 671 nm light (close to the absorption peak of the chlorophyll derivative) was better at bleaching the visual pigment than was 654 nm light, even though the latter was closer to the peak of the visual pigment alone and thus should be better at bleaching it.

The hypothesis, further developed in later papers (Douglas et al., 1999), is that the chlorophyll absorbs the long wavelength light of the red photophore and transfers the energy directly to the visual pigment, allowing the visual pigment to function as if it had a second sensitivity peak at 670 nm. In this way, the chlorophyll is known as a "photosensitizing pigment." The chlorophyll is derived from diet. Interestingly, *Malacosteus*, despite having the teeth and gape of a piscivore, mostly eats calanoid copepods, which—via their herbivory—contain chlorophyll (Sutton and Hopkins, 1996).

Finally, some researchers feel that using chlorophyll to enhance long-wavelength sensitivity may be more common than previously supposed, noting its presence in salamanders (Isayama et al., 2006). Another group found that injecting mice with the chlorophyll-derivative chlorine e_6 increased their sensitivity to red light by approximately twofold (Washington et al., 2007). Given that chlorophyll and rhodopsin are the two great absorptive molecules of life, seeing them work together to increase the visual abilities of animals in the deep-sea (where vision is hard and photosynthesis is nonexistent) is a wonderful example of evolution's economical and inventive nature. It is also a fitting way to end our discussion of absorption.

FURTHER READING

I don't know of any good and accessible books on the physics of absorption, but chapter 3 of Bohren and Clothiaux's *Fundamentals of Atmospheric Radiation* is a good if mathematical way to get started in this subject. Most optics textbooks dance around the topic.

An amazingly detailed and comprehensive book on the physical causes of color (mostly in minerals) is Kurt Nassau's *The Physics and Chemistry of Color*. Nassau is the sort of person who has probably forgotten more than you will ever know. It helps to know a reasonable amount of chemistry, and some areas of the text are heavy-going, but the general explanations are intuitive.

For biological coloration, *Animal Biochromes and Structural Colours* by Denis Fox is somewhat out of date, but full of detail on the different chemical structures and syntheses used in different contexts by animals to make color. It reads like an old school invertebrate zoology text, but about color.

The best book on the optics of vision is *Animal Eyes* by Michael Land and Dan-Eric Nilsson. This wonderful and short book gives a conceptual and comparative approach to the optics and structure of eyes. A must-read for anyone in the vision field.

There are many old, classical books on the physiology of photoreceptors and the process of vision, but I prefer a more recent publication—*Invertebrate Vision*, edited by Eric Warrant and Dan-Eric Nilsson. Even though it deals with invertebrates, the chapters cover most of what you need to know about vision.

If you are interested in the whacky world of human visual perception, Purves and Lotto's *Why We See What We Do* is fascinating. The text can be challenging if you weren't trained as a perceptual psychologist, but you can get almost everything you want from the book via the excellent figures and their captions.

The authoritative (i.e., massive) tome on color vision and perception is still *Color Science* by Gunter Wyszecki and W. S. Stiles. It's 950 pages long with not a single color figure (or even a half-tone), but discusses everything. For actual reading, I prefer *The Science of Color* edited by Steven Shevell under the auspices of the Optical Society of America.

CHAPTER FIVE

Scattering

The fog comes on little cat feet. It sits looking over harbor and
city on silent haunches and then moves on.
—CARL SANDBURG (from *Chicago Poems*)

As mentioned in the previous chapter, optics is primarily photons interacting with electrons, atoms, and molecules. If the energy of the photon matches
the difference in energy between two levels of excitation, then the photon
vanishes, its energy converted into other forms (heat, etc.). However, if the
photon's energy doesn't find a match among the differences in possible energy states, a new photon will quickly be emitted. The new photon is usually
the same energy (and thus the same wavelength) as the old one, making it
appear as if the original photon bounced, which is why this nonabsorptive
interaction is known as "scattering." It is much as if you were at the bottom
of a deep wishing well throwing a penny out every time someone threw one
in. From the viewpoint of the people above, it would look like the penny
they tossed in bounced off the bottom of the well.

When the energy of the new photon matches that of the old one (which
typically happens), the scattering is known as "elastic," just as mechanical col-
lisions that preserve kinetic energy—for example those between billiard
balls—are called "elastic." When the energy of the new photon is lower (and
its wavelength thus higher), the scattering is known as "inelastic." The photon has a roughly equal chance of being scattered in any direction. All the
observed phenomena of scattering—refraction, reflection, structural colors,
turbidity, etc.—are based on these few principles.

However, scattering is very much a "devil in the details" sort of thing, and
even simple problems can be hard to solve. Because of this, scientists since
the nineteenth century have worked hard to develop computational shortcuts. These shortcuts—Rayleigh and Rayleigh-Gans Theory, Mie Theory,
geometric optics—have unfortunately been confused with the underlying

Figure 5.1: Light passing through a transparent substance. Each location within the substance is the source of an expanding set of spherical waves, but the end result is that the light wave exiting the material on the right looks just like the wave that entered it on the left.

physics. People often talk as if Rayleigh scattering is different from Mie scattering and so on, but in reality Rayleigh scattering theory is just a set of shortcuts that give you the right answer for light scattering by small particles. You have to remember that these people did not have computers. We do. Therefore I would advise thinking of scattering as a single phenomenon that is computationally messy, rather than as a set of different phenomena. We'll get back to this idea in the next chapter.

For this and the next chapter, I will mostly consider light to be a wave. While I generally prefer to think of light as a collection of particles, interference is central to light scattering and is much more easily explained via our intuitive sense of waves. In wave language, one can model light scattering using a simple rule; divide the object into small regions, each of which emits a train of spherical waves. The height and phase of the waves depend on the intensity and phase of the incident light in that region. Every part of an illuminated object emits these waves, not just regions where the refractive index changes (figure 5.1). This may seem strange at first, but will make sense as we go on. My image of this is a pond in an advancing rainstorm. As the storm moves across the pond, each raindrop creates a circular wave that interferes with the thousands of other raindrop waves. Your job is to see what the final wave looks like at the other side of the pond.

REFRACTIVE INDEX

Before we start looking at actual particles, we need to talk about refractive index. This concept is more subtle than it first appears and we will come back to it more than once. For now, let's call a material's refractive index the ratio of the speed of light in a vacuum to the speed of light within the material. For example, water has a refractive index of 1.33, so light takes 1.33 times longer to travel through it than through the same distance in air. Since air and vacuum have the same index for all practical purposes, and biologists deal more with the former than the latter, we'll stick with air from here on.

Even this simple explanation however, immediately leads to the question of what the speed of light is. There are actually several speeds of light, each subtly different. The central definition, the all-caps SPEED OF LIGHT, is the signal velocity c, which is the speed that a light-based signal travels through air. This, so far as we know, is a fundamental property of the universe and the ultimate speed limit. It is also, according to most physicists, the speed at which light travels through all objects, regardless of refractive index. In other words, even a pulse of light traveling though a diamond, which has a refractive of 2.4 and should, according to high school physics, have a speed less than half of c, shoots through at full speed. This is easier to grasp when you remember that even dense matter is mostly empty space and that it also takes time to accelerate charges. Light is like a man running through a maternity ward, bumping into beds as he goes. By the time the babies wake up and start crying, he has already left the room. Similarly, by the time the incident light wave has started accelerating charges in matter, its wave front has already moved on. This wave front is weak but detectable. The math behind all this was developed by Arnold Sommerfeld and others at the turn of the twentieth century and gets intense (reviewed by Sommerfeld, 1954), but the basic premise is intuitive. Sommerfeld, incidentally, must rate as the most amazing academic advisor ever. Four of his students and two of his postdocs got the Nobel Prize. Sommerfeld never got one himself, which must have rankled.

What people usually think of as the speed of light is not the signal velocity described above but the phase velocity. This is the speed at which the crests of the light waves travel through the material and is directly affected by refractive index. I know this sounds a lot like the signal velocity, but it's different. First of all, the phase velocity can be negative. Second, it can be greater than c—a lot greater. For example, the refractive index of gold at 600 nm is 0.25. If the phase velocity were also the signal velocity, you could send signals through gold at four times the speed of light. Now gold absorbs light like crazy in this region of the spectrum, so you couldn't send the signal far, but going faster than c for even a tiny distance is still forbidden. The violation doesn't happen because you can't actually send a signal using a long series of identical wave crests. To do this, you need to modulate the signal, which is to say, turn it on and off or up and down. The speed at which you can do this is called the "group velocity." Interestingly, this velocity can also be negative and can also be greater than c. However, you still can't send a signal faster than light. Special relativity appears to be here to stay. The only reason I bring all this up is so that you don't get upset the first time you hear that a material has a refractive index that is less than one. Read Pendry and Smith (2004) if you would like to know more about the fascinating subject of superluminal and negative index materials.

We'll come back to what refractive index is a few more times, but for now let's think about what values it takes. For inorganic materials, refractive index can be hard to predict. Denser materials typically have higher indices, but this is only a rough guide (figure 5.2). For example, the aluminum ore cryolite is three times as dense as water but has almost the same index (which, if it's ground up into sand, makes it useful for studying digging in aquatic species since it won't scatter light under water and will thus be transparent). However, the refractive index of biological tissue components is usually linearly proportional to density with a coefficient of about 0.18 ml/g for visible light. In other words, one gram of protein dissolved in a ml of water has a refractive index of about 1.33 + 0.18 or 1.52. The increment is relatively constant from tissue to tissue because all biological molecules are made of the same atoms and bonds. There is some variation with wavelength (which is considerable once you go far into the ultraviolet or infrared) and with material, but to a good approximation the density map of a tissue correlates with its spatial variation in refractive index. This makes life simpler.

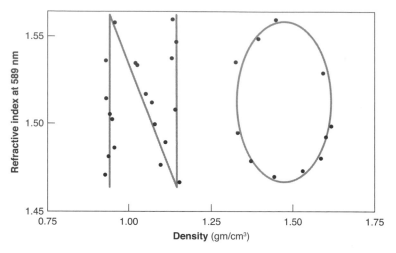

Figure 5.2: This figure, redrawn from Barr (1955) has the caption "Does the index of refraction vary directly with density?" Obviously Barr cherry-picked his data, but he does make his point.

How Light Scattering Is Measured

Before we start looking at cases of light scattering, we need to talk about how light scattering is characterized. While physicists look at this in a number of different ways, some involving nasty tensor math, the three most useful for biologists are: (1) scattering coefficient, (2) scattering cross section, and (3) asymmetry parameter. The scattering coefficient is the scattering analog of the absorption coefficient discussed in the previous chapter and is similarly easy to use. It has dimensions of inverse distance, is usually denoted by a b, and tells you the rate at which light is scattered. So, in a nonabsorbing system illuminated by a beam of light, the amount of nonscattered light still in the beam after it has gone a distance d is:

$$I = I_o e^{-bd},$$ 5.1

where I_o is the original intensity of the beam. The amount of light scattered is then just:

$$I_{sca} = I_o - I = I_o\left(1 - e^{-bd}\right).$$ 5.2

Remembering that we used a for the absorption coefficient in chapter 4, the amount of light in a beam that is neither scattered nor absorbed is:

$$I = I_o e^{-bd} e^{-ad} = I_o e^{-(a+b)d} = I_o e^{-cd}, \qquad\qquad 5.3$$

where c (which equals $a + b$) is called either the "beam attenuation coefficient" or "extinction coefficient." This term turns out to be useful when you want to figure out how something fades into the background when viewed through a turbid medium, like fog or water. In fact, almost all studies of visibility deal with c, so it's good to remember.

The problem with the scattering coefficient is that you have to measure it directly using expensive equipment. Because it usually depends on too many parameters, it is difficult to determine in other ways except for certain simple cases. It also gives you little insight into how the size and shape of the particles in the medium affect how much light is scattered and where it goes.

To better understand these two things, you need to work with scattering cross sections and phase functions. The first is fairly straightforward. Imagine a parallel beam of light aimed at a nonabsorbing particle. A fraction of this light will be scattered and thus removed from the beam. Now consider an opaque disk whose cross-sectional area is such that, if placed in the beam, it would block the same fraction of light as the particle scattered from the beam. The area of this disk is the scattering cross section of the particle (usually called "C_{sca}"). Dividing the scattering cross section by the real cross-sectional area of the particle gives another useful term, the scattering efficiency (called "Q_{sca}"). This tells you how good the particle is at scattering light.

The phase function tells you how much light is scattered in each direction. Despite its name, it tells you nothing about the phase of the light wave. This function is often normalized by the total amount of light scattered over all directions, because people are often more interested in the relative distribution of the scattered light than the absolute intensities. You can calculate the actual amount of light scattered in each direction by multiplying by the scattering cross section and the intensity of the incident light.

Probably at least 99% of biologists will never use phase functions directly, but there is a gross simplification that nevertheless nicely encapsulates how the light is scattered. Known as the "asymmetry parameter" (and oddly referred to as g), it is the cosine-weighted average of the phase function. The inverse cosine of this number roughly tells you the average angle over which the light is scattered. For example, a g of one means that all the light is scattered directly forward, a g of -1 means that all the light is scattered directly backward. A g of 0 means that as much light is scattered into the forward hemisphere as into the backward hemisphere. For a more specific example, g for the ice crystals in cirrus clouds is about 0.7. This means that about half the scattered light is within a forward cone with a half angle of 46°. Because g tells you roughly how much light is scattered forward and backward, it is used in all sorts of simple, but fairly accurate models of light propagation through dense messy substances like tissue.

While the details can get complex, only three things affect how efficiently a single isolated particle scatters light: (1) its size relative to the wavelength of light hitting it, (2) its refractive index relative to the index of surrounding medium, and—to a much lesser extent—(3) its shape. The absolute size and index of the particle don't matter. Therefore, if the relative indices are the same, a 2-cm-wide marble scatters 1-cm-wavelength microwaves just as efficiently and with the same angular spread as a 1μm wide lipid vesicle scatters 500 nm green light. So, if you can find materials with the right refractive indices, you can model light scattering in a cell using a plastic model in your microwave. Similarly, a piece of cubic zirconia (n = 2.2) in water (n = 1.33) scatters just like an identical piece of glass (n = 1.51) in air (n = 1). The absolute index of the medium does matter in an indirect way, though, because it affects the wavelength of the incident light. Size really is the critical thing, so the next three sections look at scattering from particles with diameters that are much smaller, much larger, or about the same size as the wavelength of the incident light.

SMALL PARTICLES

For particles with radii less than about 10% of the wavelength of light, all the portions of the particle experience the incident light beam as having the same phase and intensity. So they all emit scattered light more or less in phase, which means you can add up the heights of each wave scattered from a portion of the particle without worrying about wave interference. Because the energy in a wave is proportional to the square of a wave's height, scattering increases rapidly with size for small particles. Indeed, for small particles, scattering efficiency is proportional to the fourth power of the radius of the particle (relative to the wavelength):

$$Q_{sca} \propto \left(\frac{r}{\lambda}\right)^4 . \qquad\qquad 5.4$$

Since the scattering cross section C_{sca} is just the efficiency times the physical crosssection, we get the convenient equation:

$$C_{sca} = Q_{sca}\pi r^2 \propto \left(\frac{r}{\lambda}\right)^4 \pi r^2 \propto \frac{V^2}{\lambda^4} , \qquad\qquad 5.5$$

where V is the volume of the particle. So for small particles—proteins, ribosomes, etc.—scattering is proportional to the square of the volume. You have probably noticed that I have been acting as if the particle were a sphere. This is because, unless the relative refractive index is quite high (higher than what you'll find in biology), the shape of small particles has only a moderate effect on scattering.

While scattering increases rapidly with size in this range, it is still awfully small. The particles are small relative to the wavelength of the incident light and they are terrible at scattering light. The scattering efficiency of a 10-nm-diameter globular protein in visible light is only 5×10^{-7}! This means that it scatters the same amount of light as would be blocked by a disk only 0.007 nm across (that's about 14 hydrogen atoms lined up). In comparison, the scattering efficiency of a 4-μm-diameter glass bead in water is about 3.

The λ^4 term in the denominator lies behind the famous fact that small particles scatter blue light more than red light, which is at least partially responsible for our blue sky and red setting sun, which will be discussed later. Remember, it's not that blue light scatters more easily, but that a given particle looks larger relative to blue wavelengths, and larger particles scatter more light.

So how does refractive index play into all this? If you want to be exact, scattering efficiency in small particles is proportional to:

$$Q_{sca} \propto \left(\frac{m^2-1}{m^2+2}\right)^2 , \qquad\qquad 5.6$$

where m is the ratio of the refractive index of the particle and that of the medium. This is pretty clunky. However, most biologists studying scattering are looking inside tissue or at particles in water. In these cases, m is close to one and you can use the simpler equation:

$$Q_{sca} \propto (m-1)^2. \qquad\qquad 5.7$$

Even for the extreme case of pure protein particles (n = 1.55) in water, you'll only be off by about 6%–7%. You can also see from equation 5.7 that it doesn't matter whether m is greater or less than one. So a lipid droplet in cytoplasm scatters light similarly to a cytoplasm droplet in lipid. This actually holds for larger particles as well, which can help when you are thinking about vacuoles and other low-index inclusions in cells.

As I mentioned at the beginning of the chapter, small regions of matter scatter light roughly equally in all directions. Small particles do the same thing. The exact phase function looks something like a fat peanut, with the scattering in forward and backward directions being twice as much as the scattering to the sides (figure 5.3). It is described by the relatively simple equation $1 + \cos^2\theta$, where θ is the angle relative to the forward direction (we'll come back to this equation when we discuss polarization in chapter 8). So, if you are dealing with little particles, the amount of light scattered in a particular direction follows this relationship:

$$S(\theta) \propto \frac{V^2}{\lambda^4}(m-1)^2\left(1+\cos^2\theta\right). \qquad\qquad 5.8$$

This equation describes everything that matters for small particles with relative refractive indices close to one. I only left out some constants, which can be found in appendix F.

LARGE PARTICLES

For large particles, scattering is also simple. In this case, large means having a radius at least 100 times the wavelength (the exact threshold depends on the relative refractive index). So, in the case of 500 nm light, we are talking about something at least 50 μm across, like a diatom. Once you reach this size, two things happen. First, you can now use geometric optics to figure out what's

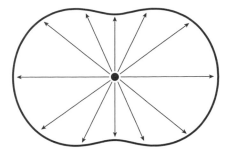

Figure 5.3: Angular scattering of light by a particle much smaller than the wavelength of the incident light (which enters from the left). Twice as much light is scattered forward and backward as is scattered at 90°. The image is rotationally symmetric around the left-right axis.

going on (ray tracing, etc.). Even better, scattering efficiency is always two. In other words, all large particles remove as much light from the beam as an opaque disk of twice the cross-sectional area. You might think that a scattering efficiency of one would make more sense, but it turns out that large particles affect the light that passes near them as well, just like waves are bent as they pass around an island. Therefore they actually remove more light from the beam than their shadow would suggest.

Of course, the actual amount of light scattered depends on the orientation of the particle relative to the light beam. A rod oriented parallel to the beam scatters much less light than one perpendicular to it, because it has a smaller cross section. But what if, as so often happens in biology, we don't know the orientation? Or have a collection of particles that are found in all orientations? In this case we would need to know the cross-sectional area of the particle averaged over all orientations A_{avg} (in other words, the average area of its shadow). In one of the great freebies of all mathematics, this turns out to have a simple answer. As long as a solid is convex (i.e., no holes or big ditches in it), its average cross-sectional area is just one-quarter of its surface area A_{surf}. So, for large particles, the scattering cross section is:

$$C_{sca} = 2 A_{avg} = 2 \frac{A_{surf}}{4} = \frac{A_{surf}}{2}.$$

5.9

Therefore, objects with a lot of surface area relative to their volume (e.g., flat disks, long rods) scatter a lot of light per volume. Spheres, which have the lowest surface area for their volume, scatter the least amount of light per volume.

MEDIUM PARTICLES

What about particles that have radii more than 1/10 the wavelength of the incident light and less than 100 times the wavelength (for 500 nm light, particles between 50 nm and 50 μm in radius)? A lot of important cellular components are in this range—nuclei, vacuoles, mitochondria, lysosomes, and so forth. The sad truth is that scattering is difficult to determine in this range, because different parts of the particle experience different phases of the incident light, and because the incident light is altered by its trip through the particle. The combination of these two effects means that the different regions of the particle are emitting scattered light at different phases and amplitudes and so a lot of complicated interference occurs. In general, scattering efficiency still increases with size, but over some size ranges, it actually decreases. When this happens, scattering actually increases with wavelength (figure 5.4). After certain volcanic eruptions and forest fires, the sky became filled with particles in this size range, and the setting sun and moon became blue instead of red, because red is scattered more than blue. Light scattering also becomes more forward-peaked with increasing size in this range, but again there are funny oscillations. The shape of particles also matters more and can be hard to predict.

The only answer to all this is to break out the computer. Gustav Mie worked out the exact solution for scattering by spheres of any size in 1908. Actually, he was beaten to it by Lorenz, who unfortunately published in Danish, which few people read at the time. Peter Debye also played a role, so to make optical physicists happy, it is best to call it Lorenz-Mie-Debye theory (see Bohren and Clothiaux, 2006, for a discussion of this). Mie, however, was the first to apply the theory to solve a real problem—the color of the famous ruby glass of Bohemia, which was due to a complex combination of absorption and scattering by 10 nm particles of embedded gold. Whatever you call it, though, the equations themselves run on for many pages and are hard to solve without a computer. The fact that a reasonable number of cases were solved in the pre-electronic calculator age is a testament to how hard people worked back then. These days, you can easily find free software packages that work with any platform. While Mie theory was designed for

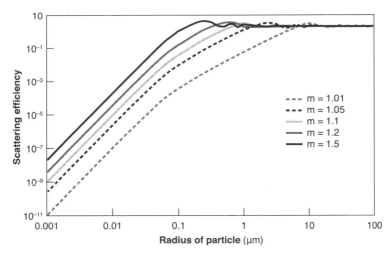

Figure 5.4: Scattering efficiency Q of spherical particles of various relative re-fractive indices as a function of their radii (wavelength of the incident light is 500 nm in a vacuum). The scattering efficiency starts low, peaks when the ra-dius of the sphere is close to the wavelength of the incident light in size, and then oscillates around a value of 2. If the medium is cytoplasm (n = 1.35), then the relative indices correspond to absolute indices of 1.36, 1.42, 1.49, 1.62, and 2.03, respectively.

spheres, it gives reasonably accurate results for any roughly spheroidal object (for example, your average organelle or phytoplankter). There are also exact solutions for flat planes and long cylinders, so, between all three, most bio-logical objects are covered. There are even rigorous solutions for concentri-cally layered spheres, which are useful for cells where the nucleus, cytoplasm, and plasma membrane all have different indices. If your particle of interest is truly bizarre in shape, you need to use finite element methods or another method of solving Maxwell's equations directly.

Of course, the downside of using a computer is that you lose any intuitive sense of what matters in scattering. A few of the patterns from small particles hold at larger sizes. For example, scattering is still proportional to $(m-1)^2$ in particles with diameters of up to two wavelengths. In general though, scat-tering cross section and the angles over which the light is scattered depend on size, shape, and index in intricate ways. For a few special cases, there are some simpler formulas developed by precalculator researchers desperate to simplify calculations, but they are not simple enough to offer any real in-sight, so you're better off with the computer.

HIDING POWER

What about refractive index? As you can see from figure 5.4, once a particle gets large enough, the amount of light it scatters is independent of its refractive index. If this doesn't bother you, it should. Surely a diamond (n = 2.4) scatters more light than glass (n = 1.5)?

It turns out that, while refractive index doesn't affect *how much* light gets scattered in large particles, it does affect *where* it gets scattered. As particles get larger, they scatter more and more of their light in a forward direction. This is due to wave interference, which we shall discuss more in the next chapter. However, high-index particles feel this effect more slowly. So how do we account for this?

In biology, you usually care about both how much *and* where light gets scattered. For example, the Small Cabbage White butterfly (*Pieris rapae*) has white wings. Electron microscopy of the scales on the white portions of the wings shows that they are covered with chitinous beads (n ≈ 1.56) 100–500 nm in diameter (Stavenga et al., 2004) (figure 5.5). The black regions of the wings have no beads, which led the authors to suspect that the beads are responsible for the white color. If so, is it because they scatter a lot of light, or because they scatter it over large angles? More precisely, if you had a volume of chitin to work with, how would you divide it up to reflect the most light? A few big beads? Many tiny ones?

The paint industry has a similar problem. White paint is white because it contains small crystals of titanium dioxide in an otherwise clear latex base. Titanium oxide has a refractive index between 2.4 and 2.6, so it scatters light well. It's expensive however, so paint developers want to find the size of crystal that scatters the most light per volume, and scatters it in all directions.

Most of the time, when something depends on two parameters, the product of both parameters is a good way to think about how to optimize it. This is no exception. The product the paint developers use is:

$$H = C_{sca}(1-g), \qquad\qquad 5.10$$

where g is the asymmetry parameter we talked about. The product H is sometimes called "angle-weighted scattering," but I like the paint industry's term for it, "hiding power" (because paint hides what's under it). Now g de-

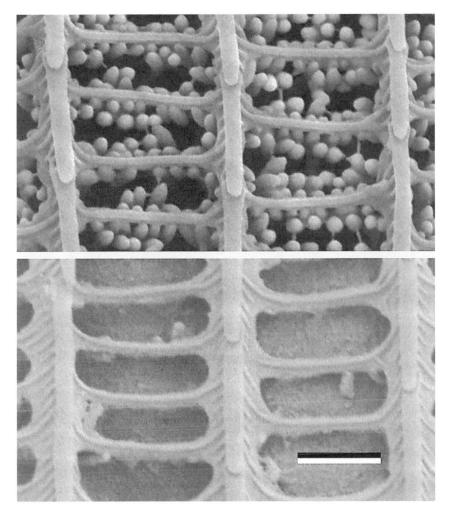

Figure 5.5: (Top) Ultrastructure of the white portions of the wings of the Cabbage White Butterfly *Pieris rapae*. (Bottom) Ultrastructure of the black portions of the wings of the same species. Scale bar is 1 mm. From Stavenga et al., 2004.

pends strongly on refractive index, so even though large particles all have the same scattering cross section, you will notice a difference in hiding power between paint made from diamond particles versus paint made from glass.

As I mentioned, usually people (and nature) care about maximizing something given a constraint, in this case volume. So what Dupont (and natural selection for the white butterfly) care about is volume-specific hiding power, or H/V. Unless your particles have a bizarre shape, this turns out to be proportional to:

$$\frac{H}{V} \propto \frac{Q_{sca}(1-g)}{r}.$$ 5.11

Large particles scatter a lot of light, but scatter most of it forward, which means that g is close to one, making 1-g small. Also, r gets large and is in the denominator, so volume-specific hiding power is low. Small particles scatter light in every direction ($g = 0$), but are incredibly inefficient, so volume-specific hiding power is again low. So where is the peak? Nicely enough, it is for particles with diameters that are between one half and equal to the wavelength of light, regardless of refractive index (figure 5.6). Break up any nonabsorbing substance into particles of this size and it will make a good white substance. Going back to the butterfly, the 100–500 nm diameter particles they have on their white wings are just the right size to scatter visible light in many directions. Two micron diameter beads would have only been half as good.

Before we got into scattering by collections of particles, I wanted to talk more about the importance of particle size. As I mentioned the volume-specific hiding power of a substance peaks when it is broken into particles with radii near the wavelength of light. On either side of this peak is a steep slope where hiding power drops by many orders of magnitude. Though the size scales are different, I like to think of this as the "crushed ice" effect. Suppose you add a certain volume of ice to a glass of water. If the ice is broken up into microscopic crystals, the total scattering will be small (you need to ignore that fact that small crystals would melt right away). If you just add two big ice cubes, you can still see through the glass fairly well. Each cube scatters a fair bit of light, but there are only two of them, so they don't scatter much light in total. However, if you crush the cubes, you can barely see through the glass at all. You have the right combination of high scattering per particle and number of particles.

This same process explains why you can see through rain, but not through fog. The volume density of water in fog is actually quite low, but if you have ever driven in it, you know it scatters light well. However, you can easily drive in a moderately heavy rain, even though the volume density of water in air in this case is far higher. This is because the volume-specific hiding power

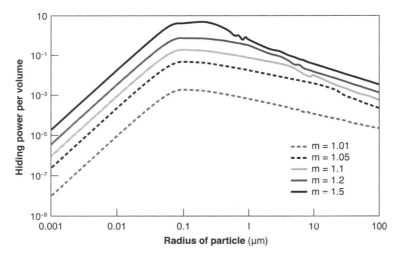

Figure 5.6: Hiding power per volume for the same parameters as in figure 5.4. Note that, regardless of refractive index, the greatest hiding power per volume is for spheres with radii of about 0.1 mm. However, the heights of the peaks depend on refractive index.

of fog-sized water droplets is much higher than for rain-sized water droplets. When I lived in Florida, we had predictable and intense thunderstorms in the afternoons. They nearly always started with a dramatic darkening of the sky, but the rain itself usually didn't appear until the sky got lighter. This is because the center of a thunderstorm is essentially a many-miles-high column of falling rain. The rain, though far denser than the clouds, scatters less light and so lets more daylight through. If you live in an area with strong thunderstorms, you can impress your friends by using the advance of lighter region to predict when the rain will start to fall.

There are also a few animals that control the size of particles within their bodies to affect light scattering. Most of these are siphonophores, but I have also seen the effect in some pteropods. All these animals while normally clear, turn white a few seconds after you touch them. It is great fun to dive with these animals, touching them to watch them decloak. The best-studied species that does this is the siphonophore *Hippopodius hippopus* (figure 5.7). George Mackie worked on these in the 1960s and found that their increased opacity was due to the precipitation of protein within their mesoglea. In their dissolved state, the proteins had a low volume-specific scattering. However, once precipitated, the larger particles had a higher volume-specific scat-

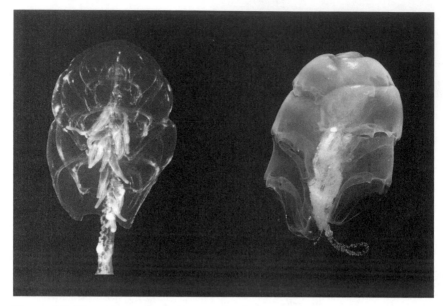

Figure 5.7: The siphonophore *Hippopodius hippopus* in its normal transparent state (left) and its opaque state (right) after being touched. Left image courtesy of the British Museum of Natural History. Right image courtesy of Edith Widder.

tering and made the animal opaque. Why these animals appear to deliberately turn opaque when disturbed, though, is anyone's guess.

Scattering by Large Collections of Particles

While it is important to understand scattering from individual particles, seldom will you ever deal with just one particle. Instead, whether working with phytoplankton in the ocean or mitochondria in the skin, you will likely be dealing with thousands to millions of them. So how does the scattering of a large assembly relate to the scattering of individual particles?

It depends a lot on how close the particles are to one another. If they are more than about ten wavelengths of light apart, then life is easy. This is because, unless they are regularly arranged, light scattered from particles separated by at least this distance interferes randomly and you have what is called "incoherent scattering." In these cases—which include rain, fog, clouds, pollution, and even dense blooms of phytoplankton—the total scattered light is just the sum of the light scattered by all the particles. Even better, the scattering coefficient b is just ρC_{sca}, where ρ is the number of particles per volume. If the particles absorb light, you can use your favorite scattering software to also calculate the absorption cross section C_{abs}. The absorption coefficient a we first talked about in chapter 4 is then ρC_{abs}. Therefore we can write equation 5.3 as:

$$I - I_o e^{-cd} = I_o e^{-\rho(C_{abs}+C_{sca})d},$$ 5.12

where c again is the beam attenuation coefficient. In other words, the microscopic meets the macroscopic, and we can directly relate the attenuation of a beam of light to the density and optical properties of individual particles.

For the moment though, let's ignore the individual contributions of scattering and absorption and think about what happens to our view of an object (i.e., its radiance) as we move away from it. Suppose we are looking at a friend of ours through fog. We know that our friend will slowly fade into the background as we move away from him or her, but how do we quantify that?

First we define the extremely useful quantity known as "contrast." The contrast of an object is simply how much brighter or darker it is than the background. However, because the eye is continually adapting to the average level of illumination, you need to do a normalization. In one form, known as "Weber contrast," you normalize by the background radiance L_b:

$$C = \frac{L_o - L_b}{L_b},$$ 5.13

where L_o is the radiance of the object. In the other form, known as "Michelson contrast," you normalize by the sum of the object and background radiances:

$$C = \frac{L_o - L_b}{L_o + L_b}.$$ 5.14

There has been much unnecessary squabbling about which definition is correct (people especially don't like that the Weber contrast of an object on a black background is infinite). This is a shame, because it is actually straightforward. If the object is much smaller than the background, then definitely use Weber contrast; if the object is similar in size to the background, then use either. Neither form of contrast is better than the other, and you can certainly use Michelson contrast for a small object or Weber contrast for small backgrounds. However, if you want to calculate the attenuation of contrast versus distance, you should use the correct one. This is because the underlying assumption for Weber contrast versus Michelson contrast is not the object's size, but whether the scattered radiance from it significantly affects the background radiance (no for Weber, yes for Michelson). Using the wrong one makes the math for contrast attenuation *much* messier. For example, the Weber contrast of your friend in the fog as a function of distance d is:

$$C = C_o e^{-cd}, \qquad 5.15$$

where C_o is the Weber contrast at zero distance. It is beautifully simple and gives you the important insight that contrast decreases exponentially with distance. Even better, the coefficient of the decrease is just the extinction coefficient. In fact, this equation is one of the most useful in biological optics, because, if you know how well an animal can detect contrast, you can use it to estimate how far away something is visible. You simply solve 5.15 for d to get:

$$d_{sighting} = \frac{1}{c} \ln\left|\frac{C_o}{C_{min}}\right|, \qquad 5.16$$

where C_{min} is the minimum contrast that the animal can detect. For humans in bright light, this is about 1% in air and 2% in water (all those scuba masks and such mess things up). This means that you can just see big black objects underwater at a distance of about $4/c$, which is a nice way of measuring c with just your eyes.

As always though, it is important to remember that c, like most things in optics, depends on wavelength. So, if you do this $4/c$ trick, ideally you should put filters over your goggles to measure it at just one wavelength. Without filters, you are integrating contrast over a broad range of wavelengths, each of which may attenuate light differently. Not a big problem for fog, but important under water where c varies a great deal over the visible range, mostly due to wavelength-dependent absorption (the wavelength dependence of underwater scattering is much smaller).

Getting back to Weber versus Michelson though, if you chose to use Michelson contrast to measure attenuation of contrast for a small object, you get the ugly and uninterpretable:

$$C = C_o e^{-cd} \frac{1}{1 - C_o(1 - e^{-cd})} . \qquad\qquad 5.17$$

However, if you are interested in the contrast between the black and white stripes on your friend's shirt, then Michelson contrast makes more sense, because there is no good reason to call either of the black or white stripes the background. The formula for attenuation in this case is not as graceful as equation 5.16, but still makes sense:

$$C = C_o \frac{1}{1 + \dfrac{L_b}{L_o}\left(e^{cd} - 1\right)} , \qquad\qquad 5.18$$

where L_o is the average brightness of the shirt at distance zero. You can see that at distance zero the contrast is C_o (as it should be) and at long distances the contrast goes to zero. How quickly it goes to zero depends on how bright the fog is compared to your friend. This makes sense, because it is the background light from the fog being scattered into the path between you and your friend that is making him or her fade away. The equation if you use Weber contrast is similar and fine to use, as long as you don't mind arbitrarily assigning one part of the pattern as the background.

You can also get around contrast and its different forms entirely by just thinking about the radiances of the object and the background. When you move away from an object (or it moves away from you), two things affect its radiance. First, the direct light from the object is scattered out of the path between it and your eye. Second, light from the background is scattered into the path (figure 5.8). This is usually called "veiling light" or "path radiance." The object light decreases and the veiling light increases with distance, so eventually the object looks just like the background. Thus:

$$L = L_o e^{-cd} + L_b\left(1 - e^{-xd}\right). \qquad\qquad 5.19$$

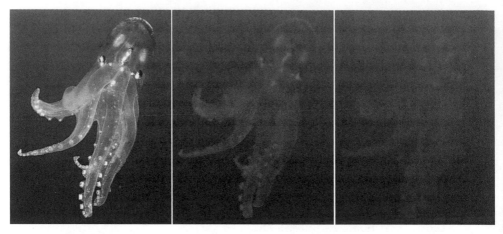

Figure 5.8: As you move away from an object in a scattering medium, the direct light from it gets attenuated exponentially and is replaced with the background light. Beyond a certain distance, known as the sighting distance, you can no longer discern the object from the background. Leftmost image courtesy of David Wrobel.

But what is the value of x, the attenuation coefficient that tells us how quickly the veiling light appears? You would think that it might be nearly impossible to determine since light is scattering in from all sides, but it turns out that x just equals c. It has to.

Imagine that your object is a patch of water, so L_o equals L_b. As you move away from the patch, its brightness should not change. After all, it's just another part of the background. The only way this can happen is if x equals c and:

$$L = L_o e^{-cd} + L_b \left(1 - e^{-cd}\right).$$ 5.20

While this looks like a lot of equations, in my experience, just three of them will solve 90% of your visibility-related problems: 5.15, 5.16, and 5.20. Before you go off and use them though, there are two remaining complicating factors, one easy, one hard. The easy one deals with the fact that the background may not stay at the same brightness as you move away from the object. On land, this usually isn't a problem. As you walk away from your friend in the fog, the brightness of the fog will likely remain fairly constant. Under water though, the world is more three-dimensional and you could be looking at your friend (now hopefully fitted with a scuba tank) from above, sideways, below, or from any angle in between. In all the previous discussion, I silently assumed that, if you were under water, you were looking horizontally so that the background radiance remained constant. Suppose instead that you were directly below your friend and looking up at him or her. You then start sinking down. Direct light from your friend is being scattered out of the path, and veiling light is being scattered in as before, but the background light is also getting darker, because you are getting deeper. As before, you might think this would be a murderously difficult problem to deal with, since the amount of background light available to be scattered into the path will change with depth, but it just involves a small correction to equation 5.15:

$$C = C_o e^{(K_L - c)d} , \qquad\qquad 5.21$$

where K_L is the attenuation coefficient of the background radiance (Mertens, 1970). If you are deeper than about 50 meters, the shape of the underwater light field doesn't change much with depth or the position of the sun, and then 5.21 is just:

$$C = C_o e^{-(c - K \cos\theta)d} , \qquad\qquad 5.22$$

where K is the attenuation coefficient of the down-welling irradiance, one of the most commonly measured parameters in ocean optics. There are giant databases with tables of c and K for multiple wavelengths and depths from all over the world. The angle θ refers to how far you are from looking directly upward. Thus, when you are looking up, the attenuation coefficient of the Weber contrast is c-K, which is smaller than c, so contrast attenuates more slowly. K is always smaller than c, but sometimes the values can be quite close. This means that, when you are looking up, you can see an object from much farther away than when you are looking down, or horizontally (assuming you start with the same contrast). You lose some range because the contrast sensitivity of eyes drops as light dims with increasing depth (try distinguishing close shades of gray under moonlight), but overall you get a big boost. Add in the fact that up is where the most light is and

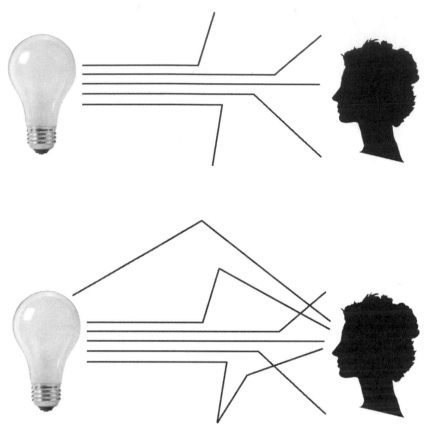

Figure 5.9: In the top image, the only light that reaches the viewer's eye is light that has never been scattered or absorbed. The rest has been scattered once and leaves the beam. The image is dim, but does not have a halo. In the bottom image, some of the light that reaches the viewer's eye has been scattered more than once. In addition, some light that originally started in another direction has now been scattered toward the viewer. Together, these multiple scatterings create a halo around the bulb because it looks like light is coming from locations other than the bulb.

that this down-welling light silhouettes anything in its path, and you can understand why natural selection has favored upward-looking eyes in deep-sea fish.

The other complicating factor is known as "multiple scattering." It truly does complicate things, and so deserves its own section.

Multiple Scattering

When a photon is absorbed, it is essentially gone, which means that the exponential absorption equation we discussed in chapter 4 works for any distance. However, a scattered photon is merely knocked out of the way, which means there is always the chance that it can be knocked back to your eye by a second scattering event, especially if absorption is low. This is what creates the halos you see around lights in fog or mist. A photon that was not originally on the way to your eye gets scattered toward it. Your eye thinks the photon originated from the second scattering position and so you see a fuzzy halo around the light (figure 5.9). If the intervening particles are all the same size and index (and thus scatter light identically), this leads to pretty effects like rainbows.

The halo can also include photons that get scattered out of the beam and later scattered back into it.

Multiple scattering makes everything much harder. Exactly how many photons get knocked toward your eyes depends on the angles over which light gets scattered and also on the absorption coefficient. The worst case scenario is something like fog, where scattering is over wide angles and absorption is low, so photons have many chances to get knocked around before being absorbed. This is why lights seen in the fog are invariably surrounded by fat halos. Under water, things are much better. The average distance a photon travels before being scattered equals $1/b$ where b is the scattering coefficient. For nearly the entire ocean, even quite close to the coast, $b < 0.1$ m^{-1}, so a photon travels 10 m before being scattered and would need to travel about 20 m to be scattered twice. Also, a scattering event may not send the photon to your eye; it could take it even farther away. Getting to your eye could take many scattering events, by which time the photon will have had a good chance of being absorbed. This is one reason that heavily absorbing waters, like tannin-laced swamps or iced tea, look so clear. Any multiply scattered light gets absorbed long before it gets to your eye. So, unless you are in murky near-shore water, you don't have to worry much about multiple scattering.

Unfortunately for biologists, tissue is far worse than fog. While it varies tremendously, the scattering coefficient in tissue is enormous. In human skin for example, b for red light is on the order of 10^4 m^{-1} (reviewed by Collier et al., 2003).The absorption coefficient also varies wildly depending on pigment content and hemoglobin concentration and oxygenation, but is generally a couple of orders of magnitude smaller (Yang et al., 2007). So, if you are studying the optics of plant or animal tissue, you *must* take multiple scattering into account. The only exceptions are tissues with exceptionally low scattering, such as lenses, corneas, mesoglea, etc. If you are in doubt, hold the tissue up to a bright, but small, light. If you see a substantial halo around the light transmitted through the tissue (or if the light is totally smeared out), then multiple scattering is occurring and needs to be dealt with.

There is good news and bad news about this. The bad news is that multiple scattering is not easy to solve. Remember that I mentioned that the equations for Mie theory fill many pages, and that you are better off buying some software? Multiple scattering is 10–100 times worse. The software you need comes in two types: (1) Monte Carlo simulations that give you the answer by shooting billions of virtual photons through your computer, and (2) radiative transfer software that solves Maxwell's equations directly using a mathematical tool called "invariant embedding." Both give the right answer. Both also have significant learning curves and neither is cheap. There are some free versions, but they are usually written for people who could probably write their own code and so they don't come with detailed instructions.

The good news is twofold. First, there is an approximate theory for multiple scattering that is simple and gives fairly accurate results. The primary restriction of the theory, developed by Kubelka and Munk (1931) to analyze paint, is that light only goes in two directions, forward and backward. While this seems like a brutal simplification, it does work.

Second, while exact numerical answers require extensive math, an intuitive grasp of multiple scattering allows you to understand a host of natural phenomena. To begin with, why is the ocean blue, and why is the blue color more saturated in clearer waters? Originally it was thought that the ocean's color was simply a reflectance of the sky's color. Having been to sea many times, though, I can tell you that the ocean still looks quite blue on an overcast day, sometimes even bluer. So is it due to scattering? We know that scattering from small particles is inversely proportional to the fourth power of wavelength, and particles in the ocean seem small, so possibly the ocean is blue for the same reason that the sky is blue. However, actual measurements of scattering in the ocean show that its wavelength dependence is fairly weak—not enough to create the deep blue color we see.

This leaves absorption. As I have mentioned several times, water is a selective absorber of light and most transparent to blue light. However, if the ocean didn't also scatter light, then it would look black from above (though it would look blue if you were under water looking up). This is exactly like the Yunzi stones I described in chapter 4. Without scattering, even heavily pigmented objects look black, unless you view them via transmitted light. With the exception of some bioluminescent animals, there are no underwater sources of light, so a scatterless ocean is black in nearly every direction.

For the ocean to look blue from above, we need both absorption and scattering. The blue color we see from a ship is due to photons that entered the ocean and then were scattered so many times that they turned 180 degrees and came back out. How many times? Quite a few, if the water is clear, like it is once you get off the continental shelf. Out there, the scattering coefficient is about 0.1 m⁻¹, so a photon goes about 10 m before it can change direction. The asymmetry parameter g is about 0.9 (Mobley, 1994), which roughly means that the photon, on average, changes direction by $\cos^{-1}(0.9)$ or 25 degrees. So, even if every scattering event is in the same direction and plane, the photon has to be scattered seven times to completely turn around and exit the water. These seven scattering events will take it through an average of 70 m of water, a trip that only blue photons will survive. Nearly all the photons of other wavelengths will get absorbed before they make it back out. In reality, some of the scattering events will cancel each other out, leading to a random walk that can be long (to give you an idea of the inefficiency of random walks, it takes thousands of years for a photon generated in the core of the sun to make it to the sun's surface, but then only 8 minutes to get the rest of the 93 million miles to us). I often scuba dive over water that is extremely clear and thousands of meters deep. Looking down at the up-welling light, I can't help but wonder how far some of those photons traveled to get back up to me.

What if there was a bloom of coccolithophores in our patch of ocean? Coccolithophores are a type of phytoplankton that produces (and later sheds) calcite scales that are about 1 μm to 10 μm in diameter. The scales absorb almost no light but, due to their size and high index relative to water, they scatter light like crazy, driving the scattering coefficient up to about 3 m⁻¹ (Balch et al., 1991). The asymmetry parameter stays about the same, so the average trip a photon will need to take before it is turned around will be $3/0.1 = 30$ times shorter. This shorter trip means that more nonblue pho-

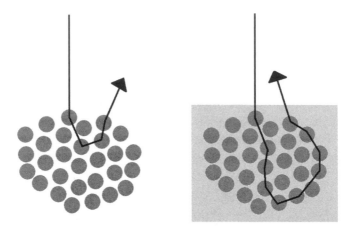

Figure 5.10: Light scattered by dry sand (left) is spread over a wide range of angles and thus exits the sand before much light has been absorbed. Light scattered by wet sand (right) is spread over a lower range of angles because the relative index of the sand compared to water is lower than sand versus air. Thus, the light travels a longer path through the sand and more of it gets absorbed, making the wet sand appear darker.

tons will survive and that the ocean's color will be less pure. In fact, oceans during coccolithophore blooms are a milky and pale blue.

A similar thing happens if you change the angles over which light is scattered. If the light gets scattered in a more forward direction, it takes even longer for it to get back out of the material, leading to many more chances for absorption. Craig Bohren invoked this principle to explain why many wet things are dark. The relative darkness of wet objects like sand and cloth is so common that most of us don't even bother to think about why it happens. Suppose you are walking along the beach near the water on dry sand. Light from the sun and sky enters the sand and is scattered. The difference between the refractive indices of sand and air is fairly large, so light is scattered over large angles and exits the sand relatively quickly. This gives the light relatively little chance of being absorbed by the pigments in the sand grains and so the sand looks light. Now suppose a wave washes over the sand. The water sinks in, replacing the air. The water has a higher index than the air and so the difference between it and the index of the sand is smaller. Thus the light is scattered in a more forward direction, takes longer to get back out, and has a greater chance of being absorbed (figure 5.10). So the wet sand is darker.

One way to prove this to yourself is to pour a high-index fluid, like oil, onto sand. You will see that the sand gets even darker than it does with water. It's also more colorful, because, as we discussed in chapter 4, an increase in absorption also increases the wavelength dependence of the absorption and so colors get more saturated.

BACK TO REFRACTIVE INDEX

Because refractive index affects scattering and because scattering can in some ways be considered a failed absorption, you might suspect that refractive index and absorption coefficient are related. And indeed they are. Refractive index for physicists is actually a complex number (in the mathematical sense). The real part of this number (n) is what we as biologists usually think of as refractive index, using an unspoken and often unrealized shorthand that for the most part works because few biological materials are strong absorbers of light. The imaginary part of the refractive index (k) is closely related to the absorption coefficient a. I.e.

$$n_{complex} = n - ik, \text{ where } n \text{ is "refractive index" and } k = a\frac{\lambda}{4\pi} \qquad 5.23$$

Both the real and imaginary components of the refractive index depend on wavelength via what are known as the "Kramers-Kronig Relations," which represent the wavelengths where absorption occurs as resonance peaks. Few biologists will ever need to use these relations, but it is good to know that refractive index and absorption coefficient are in some ways two sides of the same coin.

Figure 5.11 shows the real and imaginary refractive index of water over a tremendous range of wavelengths. The first thing to notice is the giant drop in absorption in the visible range. As I have mentioned, if it weren't for this narrow window, we would all be blind. The second thing to notice is that absorption peaks are often correlated with oscillations in refractive index (it is difficult to see, because the scales in the graphs are different and cover a huge range). This shows you how tightly coupled the two parameters are. You can also see this unity by looking at reflection from a smooth surface. You have noticed that clear substances with high refractive indices (relative to the medium they are in) reflect light. Glass is a good example. However, the equations for this, known as the "Fresnel-Arago Equations," don't care

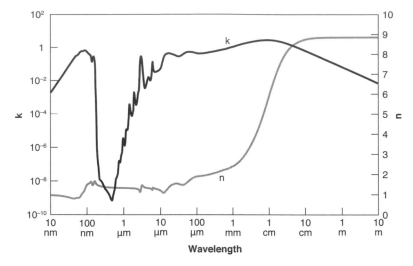

Figure 5.11: Real (n) and imaginary (k) components of the refractive index of water. The real component is what we usually think of as the refractive index; the imaginary component is closely related to the absorption coefficient. Note the gigantic drop in k in the visible range. Also notice that the real refractive index varies by nearly a factor of ten over the range shown. Data from Segelstein, 1981.

whether it's the real or the imaginary component of the substance that is high. In other words, a substance with the same real refractive index as the medium, but a higher absorption (which corresponds to imaginary index) will also reflect light. Similarly, a particle that matches the real index of a medium but has a different imaginary index will still scatter light. Therefore, you cannot say that imaginary index only causes absorption. Unfortunately, the good examples of scattering via imaginary index changes are all metals and other nonbiological substances. Overall, the connection between what we call refractive index and absorption coefficient is not strictly relevant to most biological research, but seeing hidden unity is clarifying and good for the soul.

More relevant to biologists is the fact that light doesn't "see" refractive index changes that happen over small size scales. A good rule of thumb is that any refractive index variations that are on size scales less than one half a wavelength across have no effect on the transmission of light. In other words, something can be incredibly complex at small size scales, but as long as it looks homogeneous at size scales greater than 250 nm (for 500 nm light) it will look clear. You might be thinking, "But we just spent the beginning of this chapter talking about scattering from particles that are far smaller than that." Yes we did, but we were talking about *isolated* particles. Even a 50 nm particle affects what happens on larger size scales if it sits on an empty background.

Those of you who have done microscopy may recognize the relationship between this and the resolution limit of a microscope. The best microscope can only resolve objects down to half a wavelength of light, but, as you may know, it can see an object smaller than that if it is on an empty background. In graduate school, a lab down the hall studied microtubules. These are only about 25 nm across, but you can see them with a microscope. Their width is blown up to the resolution limit of the microscope and you can't separate two of them if they are closer than about 250 nm apart, but you can certainly see them.

Nothing is perfectly homogeneous. If nothing else, you eventually get down to individual atoms. So it is good that light isn't affected much by variations on scales less than half its wavelength. Otherwise, nothing would ever transmit light and again we would all be blind.

So if the variation in refractive index at these size scales isn't seen, then what is the refractive index at larger size scales? It turns out to be just the volume-weighted average of the index of the components. For example, a spongelike composite of 30% air and 70% chitin that is homogeneous at size scales greater than 250 nm behaves more or less like a solid substance with a refractive index of 0.3*1 + 0.7*1.55 = 1.385.

This fact has been used by animals to perform a couple of neat tricks. The first is found in human lenses. Lenses are composed of elongated, enucleated cells that are essentially bags of high-density protein. As such, they scatter little light. Unfortunately for them, their membranes have a higher refractive index, and so light can scatter at the membrane-cytoplasm interface. The difference is small, and the scattering is low, but there are a lot of interfaces along the path through a lens. Without any compensation for this, the lens would be opaque. One solution is to add proteins to the membrane so that it more closely matches the index of the cytoplasm. There is some evidence for this. Another solution is to have a space between neighboring cells that has a lower index than that of the cytoplasm. If the space is just the right thickness, the volume-weighted average of the indices of the membrane and the space will match the index of the cytoplasm and the membrane interface will no longer be seen by the light (i.e., there won't be any scattering). It looks as if this trick may be going on in human eyes. Interestingly, many cataracts show an increase or decrease in the width of this extracellular space, which would disrupt this fine balance and lead to increased scattering (Costello et al., 2010).

But what about cases where light is going from one medium to another, each of which is thick—for example, air to the chitinous cuticle of a beetle? No sort of averaging trick is going to make that interface disappear, and so light will be reflected even if the surface is transparent (we will discuss the details of this in the next chapter). It's not much light, only about 4% if the light hits the surface perpendicularly, but this can be enough to make an otherwise hidden surface visible. Think about the reflections you see from windshields and windows on a sunny day. It also reduces the amount of light that gets transmitted, which is a problem if the interface happens to be the cornea of an eye. At more grazing angles of incidence, the fraction of light gets much higher. For example if the light hits the surface at 45°, about 12% of the light is reflected. So a solution would be desirable.

The designers of camera and microscope lenses are especially aware of this issue. It turns out that the best way to make a good, aberration-free lens is to compose it of many weak lenses in series. The individual lenses, due to their varying shapes and optical properties, balance one another's imperfections, creating a lens system that is nearly perfect. Thus, many camera lenses are made of over ten lenses. Microscope objectives are even worse, a few having over twenty lenses. The difficult process of making these tiny lens elements and assembling them in a perfect line is a major reason why these things are so appallingly expensive. Another big issue is that these systems have many air-glass interfaces. While each only reflects a small percentage of the light, a system of ten air-spaced lenses has twenty interfaces. Even if each interface only reflected 4% of the light, the total amount of light that would get through the lens without suffering at least one reflection is about 44% (= [1-0.04]^20). The remaining 56% of the light is reflected around inside the lens, leading to all sorts of flare, glare, and ghost images.

The human solution to this is coating the lens with thin films of substances with particular refractive indices. Doing this well has become an entire field of research, but the simplest solution is to use one layer, a quarter of a wavelength thick that has a refractive index that is the geometric mean of the index of the medium and the substance, in this case air and glass (i.e., $\sqrt{n_{air} n_{glass}} = 1.23$). Magnesium fluoride is used since it is durable and has about the right index. However, this only works for one wavelength, so modern lenses have multiple coatings that are determined after much modeling. These can bring the reflection from a surface down from 4% to less than 0.5%.

As far as we can tell, no animal uses thin film coatings to reduce reflection. Instead, capitalizing on the fact that nature is better at making small structures than she is at using substances with unusual refractive indices, we see what have been called "corneal nipple arrays" or "moth eye structures." These are seen on the transparent wings of some moths and (as you might guess) the eyes of moths (Stavenga et al., 2006). Under a scanning electron microscope, they look like a series of bumps with widths far less than half a wavelength of light. If they were about ten times larger, the bumps would scatter light like mad and make the wing or eye white and opaque. However, at this size scale what they do is change the refractive index that the incident light "sees" as it approaches the main surface of the wing or eye. When the light first encounters the bumps, it only sees the peaks, which only take up a small fraction of the area at that height from the surface. So the average refractive index at that height is close to that of the medium, in this case air.

As the light moves closer to the main surface, the bumps occupy an increasingly larger fraction of the area, so the average index grows larger, eventually reaching the value of the surface (in this case chitin) (figure 5.12). Therefore, what was a sharp jump from a refractive index of 1 to one of 1.5 is now a smooth gradient that reflects far less light. If the shapes of the bumps are just right (known in the microwave world as the Klopfenstein taper), the reflection can go all the way down to zero. However, even your average bumps work fairly well.

It is rumored that the B2 stealth bomber has a nipple array coating that works at radar frequencies, but I have never been close enough to one to know for sure. A Swedish company also makes plastic nipple array coatings that you can put on large windows and glass doors to increase light transmission. Having spent time in Scandinavia, I can see why you would want every last photon. In the case of the moth eyes, it is assumed that the structures on the eyes increase the transmission of light and thus the sensitivity of the eyes. The difference would be small, though, especially compared to the enormous

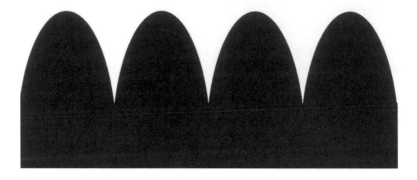

Figure 5.12: (Top) Side view of a moth eye structure, showing small bumps of chitin (in black: index = 1.5) in air (index = 1). (Bottom) Photon's "view" of the same surface— the refractive index changes smoothly from 1 to 1.5 as more and more of the area at any given distance from surface is taken up by the chitin.

dynamic range of daylight and the logarithmic sensitivity of eyes. Conversely, it could help to camouflage the eyes, which are large and could give away the animal via reflection in bright light. Or the arrays could serve no significant optical function at all, aside from their inspiration for human technology. Similarly little has been proven about the arrays on the wings of some moths. They do reduce reflection, which makes the animal more cryptic, but the actual structures are the bases of the (missing) scales, which would be there anyway. As usual, this all raises more questions than it answers. It would be particularly interesting to look for nipple arrays in transparent, pelagic animals. The reflection is less, since the animal is sitting in water, which has a higher index, but can still give the animal away. Given the importance of transparency as a form of camouflage for certain pelagic species, anything that furthers it would be advantageous. I would be especially interested to see if any of the deeper transparent species have these arrays, since they are facing bioluminescent searchlights that pick up any reflection at all.

FURTHER READING

There are three bibles of light scattering, all challenging to read unless you have a fair bit of background in math and physics. They cover the same material, but with different flavors and emphases. The oldest and my personal favorite is H. C. Van de Hulst's *Light Scattering by Small Particles*. It was written before computers made life simple and can give you an idea of how clever people were back then. The most recent is *Absorption and Scattering of Light by Small Particles* by Craig Bohren and Donald Huffman (of Buckeyball fame). It is an excellent and comprehensive text that humorously examines a lot of dogma and missteps, but it is challenging. Some biologists, including one of my graduate students, prefer Milton Kerker's *The Scattering of Light and Other Electromagnetic Radiation*, which is written more from a chemist's point of view and is particularly good at discussing scattering from inhomogeneous solids and dense liquids, both useful for a biologist.

An excellent book on underwater visibility is Lawrence Mertens deceptively titled *In-Water Photography*. While it sounds like a coffee-table book for scuba shutterbugs, it is actually an excellent and clearly written treatment of how things look under water. I have often wondered what the real story behind this book is, especially since chapter 5's discussion of supplemental lighting invokes nuclear power sources, including those based on plutonium.

All the original work on underwater turbidity and visibility was done in Seibert Q. Duntley's Visibility Laboratory at Scripps. So far as I know, a book was never published, but your local academic library can find copies of the voluminous government reports they wrote. Just as everything in anatomy was first discovered by a nineteenth-century German biologist with a hand lens and good drawing skills, almost every idea in visibility and turbidity was first discussed by Duntley's group in the 1940s and 1950s.

Very few popular discussions of light scattering exist. Bohren's previously mentioned *Clouds in a Glass of Beer* and *What Light Through Yonder Window Breaks?* have some of the best examples. Chapters 14–22 in the former are especially good and form the basis for some of the discussions in this chapter.

CHAPTER SIX

Scattering with Interference

> Reflection and transmission are really the result of an electron
> picking up a photon, "scratching its head," so to speak, and
> emitting a new photon.
> —RICHARD FEYNMAN (from *QED*)

Light does not bend in a lens, it doesn't bounce off the surface of glass, and it doesn't spread out after passing through a small hole. It doesn't even travel in a straight line. The happiest day of my scientific life came when I read Feynman's *QED* and learned that refraction, reflection, and diffraction—things I had known since fifth grade—were all lies. More accurately, they are illusions. It *appears* that light bends, bounces, and spreads out. The illusion is so good that you can base solid mathematical predictions on them, but careful thought and further experiments show that more is going on.

So where do all the effects we see—reflection, refraction, colors in soap bubbles—come from? They come from the fact that photons interact with one another in an unusual way. The interaction is often referred to as wave interference because it can be described mathematically much like the interference of water waves, but the reality of it is considerably stranger. The strangeness is discussed in more detail in the last chapter, but a hint of it can be seen from the fact that photons appear to interfere with one another even if they are emitted one at a time, which water waves obviously don't do. For this chapter and the following one, however, it is best to think of scattered photons as expanding spherical waves of electrical fields with amplitude (field strength) and phase (where the waves are in their cycle of changing electric field), one coming from each small part of the material that the light is going through (see *QED* for a neat explanation of all this in photon language).

Before we start, I want to make clear that I am presenting all this in simplified form. Dotting the *i*'s and crossing the *t*'s involves complex analysis. In this case "complex" refers to complex numbers, though it can be complex in the usual sense as well. My goal here is to present the basic ideas so that you can see the unity underlying a wide range of phenomena and further appreciate some of the tricks that animals do with light.

INTERFERENCE IN A VERY SMALL NUTSHELL

Before this chapter (and for much of the later chapters), we added light like we add most things— $1+1=2$. More specifically, if the irradiances of two monochromatic beams of light striking the same spot were both A, the total irradiance was $2A$. However, the waves within each beam have phase. If you take these phases into account, the total *instantaneous* electric field at the spot where the waves meet is:

$$E = a\cos(t) + a\cos(t + \phi),$$ 6.1

where a is the amplitude of each wave (the strongest the electric field can get), t is time, and ϕ is the phase difference between the two waves. For clarity, I have set the frequency of both waves to one and set the time so that the phase of the first wave is zero at time zero. I have also assumed that the electric field is a scalar instead of a vector. All this affects the details of the calculations, but not the general arguments I am making.

No eye or real device can measure the instantaneous electric field at visible light frequencies. Instead, the irradiance we see/detect is proportional to the average of the *square* of the total electric field over many cycles. I promised to avoid too much math, so I am going to skip the messy integral that gives you the averaged irradiance in this case. Instead, I would like you to notice what happens to the instantaneous electric field for certain values of the relative phase difference ϕ. If the phase difference is 0, 360°, or any multiple of 360°, then the intensity of the sum of the two electric fields is always double the intensity of each electric field. Specifically, this means that the summed amplitude (the highest the electric field gets over one cycle) is also double that of each wave. Since the irradiance you detect is proportional to the square of this amplitude, the combination has four times the irradiance. In other words, $1+1=4$. This is called "constructive interference." If the

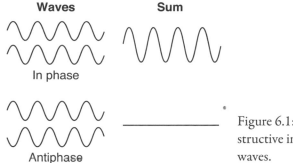

Waves   **Sum**

In phase

Antiphase

Figure 6.1: Constructive and destructive interference of simple sine waves.

phase difference is 180°, or 180° more than any multiple of 360° (e.g., 540°, 900°), then the value of the electric field of one wave is always the exact opposite of the value of the other wave and thus their sum is always zero. So you detect no light at all and 1 + 1 = 0. This is called "destructive interference."

What about the other possible phase differences? As you go from a phase difference of 0° to 180° (with or without multiples of 360° added on), the measured irradiance drops smoothly from four times the irradiance of either beam to 0. From 180° to 360°, the measured irradiance goes back up to four times the original irradiance again (figure 6.1). If two beams start out with the same phase (for example, one laser beam made into two via a beam splitter), then their relative phases when they meet again (assuming they do) depends on how far they each traveled. It also depends on what the two beams traveled through, since, as we discussed in chapter 2, the wavelength of a beam of light is affected by the refractive index of the medium in which it travels. This is wave interference in a nutshell. Note that neither constructive nor destructive interference makes or destroys light, which would be a violation of conservation of energy. Instead, you can think of interference as changing the locations where light is detected. The total amount of light stays the same.

There is a good chance you have seen interference fringes in a high school or college physics lab exercise, usually made by a laser beam going through a small vertical slit. You may have also been told that the colors you see from oil slicks, soap bubbles, and iridescent feathers and butterfly wings are due to interference. While true, this leaves most people with the impression that interference is a limited phenomenon, found only when monochromatic light interacts with special structures. However, interference is at the heart of many important phenomena. Without it, lenses wouldn't focus, window glass and still ponds wouldn't reflect light, and the ocean and indeed the inside of our eyes would be as opaque as milk.

DOES LIGHT TRAVEL IN A STRAIGHT LINE? TRANSPARENCY

In the previous chapter, we talked about light traveling through water and pretended that nearly all the scattering was due to the particles in the water. This is commonly done, leaving people to assume that light doesn't interact with transparent substances like water and glass. I grew up believing that glass and water were transparent because the photons just weren't interested in the molecules that made up these substances and shot on through like late-night express trains.

Nothing could be further from the truth. While water and clear glass don't strongly absorb visible light, they scatter it like mad. Even in the purest water, a photon travels only a microscopic distance before being scattered by a molecule. In glass, it's even worse. If each scattered wave were added up without considering phase, even small thicknesses of both substances would be opaque.

So why do water and glass appear clear to us? The reasoning is the same for water and glass, so let's choose water and go night diving. Suppose your dive buddy is shining a flashlight toward you. Suppose also that you are in the open ocean and the water is extremely clear. The emitted light has created a tremendous number of expanding wavefronts of scattered light in the water between you and your buddy (figure 6.2). In fact, a wavefront is being emitted from roughly every point in the intervening water. So, what would a third diver, floating above both of you, see? Would he see the beam of light traveling from your buddy to you?

No, and here's why. Suppose the third diver is looking at one spot along the path between you and your buddy. The light he sees is the averaged square of the sum of all the electric fields of all the little scattered waves that were created by the flashlight beam along the path. The sum of these countless fields depends on the phase of each one, which depends on the distance each particular wave has traveled from the beam path to the third diver. Because the water molecules are so close to one another, the spherical waves emitted from each one have all possible phases when they arrive at the third

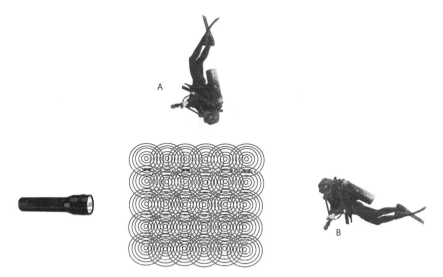

Figure 6.2: Light from a dive flashlight travels through the water, and is scattered all along its path. However, assuming the water is clear, diver A does not see the beam due to destructive interference, but diver B does.

diver. Thus, these uncounted numbers of waves cancel the others out at the third diver's eye (much like a very large set of random numbers between −1 and 1 add to zero), and he sees nothing. Now, water is never perfectly clear. Also, random thermal fluctuations of the water molecules themselves create larger regions of higher and lower density that lead to observable scattering, but we are ignoring that for now. The important thing is that, due to the summing of large numbers of paired waves of opposite random phase, a substance can appear to not interact with light at all, even though it scatters light at every location. To me, this is the ultimate sleight of hand.

But what about light in the forward direction? Doesn't the previous argument mean that *you* won't see light from your buddy either? You will, though, because in this case all the scattered light is in phase. The reason for this is simple. Imagine, in a super slo-mo world, that the light wave from the flashlight has reached the first molecule of water. As we discussed in chapter 4, the electric field in the light wave starts oscillations in the molecule that then induce another light wave of the same frequency that now expands in all directions. However, it takes the light wave a little time to accelerate the charges in the water molecule that emit the scattered wave. Even electrons take some time to accelerate. Due to this delay in acceleration, the forward-scattered wave has a definite phase relationship with the main wave, like a kid walking exactly three steps behind its mother. Both the main light wave and the scattered wave move forward to the next molecule. The main wave is much more intense than the first scattered one, so we can ignore the second's effect on the next molecule. Again, after a short delay, the main wave induces a scattered wave in the second molecule. Because both the first and second molecule experienced the same delay (the delay depends on the substance, which is all water) and because the scattered waves have this defined relationship with the main wave, the two scattered waves are in phase with each other. These two scattered waves and the main wave move on to the third molecule, and so on, until eventually the main wave is trailing a clutch of scattered waves like baby ducks, except that, instead of being in a line, they are all in phase. This means they interfere constructively, and you, as the diver viewing the flashlight, see a bright light.

The delay we just mentioned also implies that the scattered waves, while all in phase with one another in the forward direction, are not in phase with the main incident wave. The observable light wave traveling through the water is the sum of these two waves, taking interference into account. We won't do the math here, but the observable wave has a shorter wavelength than the original wave. Since it still has the same frequency (and it always will unless energy is added or taken away), this means that the phase velocity is slower. You guessed it—this delay is what is responsible for the refractive index of a material. Some materials have odd delays that lead to summed waves that have a phase velocity faster than that of light, and some summed waves even go backward (or backward faster than the speed of light). As I discussed in chapter 5, none of this violates Relativity—you still can't send a signal faster than light. The main wave and the scattered wave continue to travel at the usual speed c; it's just their sum that looks like it's doing weird things.

The bottom line though is that there is a difference between scattering and *observable* scattering. Transparent objects do scatter a lot of light, but only the forward scattered light is observable (or at least it's by far the most dominant). The interference between this forward-scattered light and the main beam is what changes the wavelength and phase velocity of the detected beam, leading to what we call refractive index.

Transparency is of course critical to the oceans and other aquatic habitats. In fact, life, if it could have existed at all, would have gone in a very different direction if light and other radiation couldn't make it more than a micro-scopic distance under water (actually it wouldn't have even made it through the lower atmosphere). Transparency is also important for lenses and eyes, which cannot function if they scatter massive amounts of light. Transpar-ency is also near and dear to my heart because it is found in many of the pe-lagic species I study. The pelagic environment, especially in the open ocean, is a frightfully bad place to be if you are a prey item. It's featureless, with only the gentlest gradients of color and brightness. In addition, if you are discov-ered, there is no nearby hole to dive into or rock to jump behind. So, unless you are fast, large, or protected by other defenses, the game is over once you are spotted by a predator. This has led to the evolution of some impressive camouflage tactics rarely seen in other environments, including transpar-ency, which is found in the pelagic members of nearly every major taxonomic group (reviewed by Johnsen, 2001) (Plate 3). The adaptation is beautiful, with many of the animals looking like cut crystal, and—at least in some cases—is likely to be challenging to achieve. Some, like many jellyfish and comb jellies, are nearly 100% water and so are transparent for fairly trivial reasons. Others, like the larvae of eels and spiny lobsters, are transparent be-cause they are thin. Remember from chapter 5 that light absorption and scattering are exponential, so anything can be clear if it is thin enough, even metal. However, there are many pelagic animals that are not flat and have dense tissue but are still clear, including certain fish, cephalopods, crusta-ceans, and worms. How they manage this is still poorly understood. From histology and the fact that the animals become opaque soon after death, we know that it is not a simple case of being mostly water.

We also know that the transparent tissues that have been studied in de-tail—human lenses and corneas—are heavily modified. Lenses consist of long cells that are enucleated and contain few to no other organelles (the rare exceptions hide behind the iris where they have no effect on vision). Instead the cells are filled with high concentrations of a handful of different pro-teins, all called "crystallins," though they are not crystals or even the same protein type from species to species (usually they are a housekeeping protein with good resistance to denaturation and clumping). This high concentra-tion gives the lens both its high refractive index and, by squishing all the proteins together, enforces an order that makes the lens transparent in the same way that water and glass are. Since the crystallins are free to move, if the concentration were lower you would have regions where the concentration would be higher or lower than average at any given moment, leading to ob-

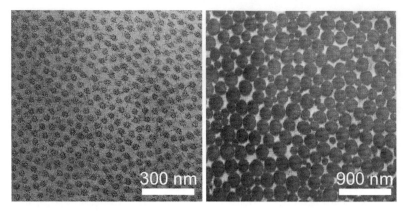

Figure 6.3: Electron micrographs of the cornea (left) and sclera (right) of a human eye. Note that the collagen fibers in the cornea are smaller and more regularly arranged. From Vaezy and Clark, 1994.

servable scattering. Alison Sweeney, explored this in her doctoral thesis and found that crystallins in low concentrations may actually electrostatically repel one another to maintain order (Sweeney et al., 2007).

Corneas are transparent because they are highly-ordered arrays of collagen fibers that are transparent in the same way as water and glass. A good rule of thumb, which both lenses and corneas follow (and that we mentioned in chapter 5), is that, for a substance to be clear, it has to look uniform at all size scales larger than half a wavelength of the incident light (Benedek, 1971). The white of the eye is made of the exact same fibers as those in the cornea, but those in the former are larger and randomly oriented, leading to observable scattering and thus opacity (Figure 6.3).

How transparent animals become so and remain so is not yet known, but microscopic studies of some of them do not show obvious differences from other animals in morphology or ultrastructure. We have found some subtle changes that should reduce scattering (Johnsen and Kier, unpublished data), but it is not known if these are enough to account for their transparency. It is possible that certain transparent species use a "clearing agent," essentially any sort of innocuous substance that raises the refractive index of the cytoplasm and extracellular fluid so that it is closer to that of the higher index components, such as dense concentrations of protein, lipids, and nucleic acids. Again, this is a fascinating field with more questions than answers.

Before we leave biological transparency, there are two final wrinkles. The first is that transparent animals need to eat. It turns out that, for the most

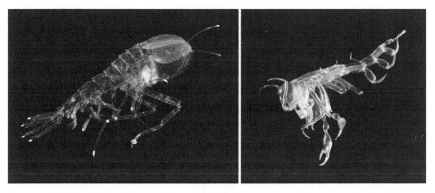

Figure 6.4: (Left) The hyperiid amphipod *Cystisoma* sp. Its eyes, which fill its head, have a thin retina right below the lenses of the facets. (Right) The hyperiid amphipod *Phronima* sp., whose tiny retinas (dark area on head) are connected to the large lens array at the top of the head by tapered fiber optic cables. Images courtesy of Edith Widder.

part, even transparent animals become opaque when they are dead and are being digested. In addition, as we discussed in chapter 3, many oceanic animals are bioluminescent and flash like mad while being chewed and digested. So you need an opaque gut that you can also hide. Certain squid, heteropods, and hyperiid amphipods have dealt with this is in a clever way. They have a needle-shaped gut with attached muscles that ensure that, no matter where the animal is pointing, the gut is vertical. This minimizes the silhouette of the gut for animals looking up from below or down from above (Seapy and Young, 1986). They deal with the side view by coating the sides of the gut with mirrors, which we'll get to later in the chapter.

The second wrinkle is that many transparent animals also like to see. Retinas, by definition, absorb light and so cast a shadow. This is especially a problem for deeper animals that need large eyes and highly absorbing retinas that catch every photon. Hyperiid amphipods—big crustaceans that parasitize jellyfish—have dealt with this in two ways (Figure 6.4). One amphipod, *Cystisoma*, has gigantic compound eyes. In fact the whole head is just the two eyes. However, the photoreceptors are directly behind the corneas of each facet of the eye, spreading out the retina to a point where it casts a weak shadow. In contrast, the hyperiid *Phronima* (the inspiration for *Alien* for its sinister appearance and nasty parasitic method of reproduction) has four large sets of facets that are connected with four tiny retinas. I say "connected with" rather than "focused on" because each facet of the eye is connected to a set of photoreceptor cells via a tapered fiber optic cable. This is necessary because the retinas are too small and clumpy to receive an image via typical geometric focusing. This remarkable adaptation allows the animal to have the large optical apparatus required for excellent low-light vision and to keep the tiny retinas that cast almost no shadow (Nilsson, 1982).

DOES LIGHT BOUNCE? REFLECTION

Another thing most of us learn early on is that light bounces off of certain materials and bends when it enters others. Just as Rayleigh and Mie scattering are different mathematical solutions for the same process, reflection and refraction are both special and mathematically tractable cases of scattering. In other words, in certain configurations, scattering can be described by the far simpler equations for what we call reflection and refraction. However, the equations have been around so long that they have turned into laws. I suppose that the line between an equation and the process it describes is a fine one, but in this case it has led to the subdivision of phenomena that are simply special cases of one underlying process.

Let's start with reflection and for the moment not worry about what the light is bouncing off of. Probably the first optical fact I learned was that the angle of incidence equals the angle of reflection. No doubt, this is a useful thing to know, but does light really behave this way? If you are looking at a light bulb in a mirror, is it true that the light you see somehow works out the correct path so that the angle of incidence equals the angle of reflection? If so, why?

This is another case where destructive and constructive interference play a major role. As you might guess, light from the bulb can bounce off the mirror and reach your eye by any number of paths (ignore scattering by the air) (figure 6.5). These paths have different lengths and so the light that reaches us arrives with many different phases. The difference in distances is so large and continuously distributed that, as with the water we discussed above, nearly all phases are present and the amplitudes cancel one another out. However, when the angle of incidence equals the angle of reflection, something special happens. For this particular set of angles, the total distance from bulb to eye via the mirror is minimal. More importantly, the *change* in

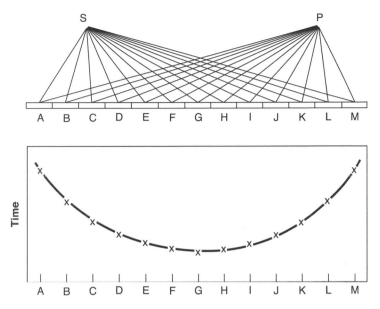

Figure 6.5: Light leaves the point *S* and bounces off all parts of the mirror. For most of the parts of the mirror, the time it takes for the reflected light to reach point *P* varies quickly with location, leading to destructive interference and no observable light. However, when the angle of incidence equals the angle of reflection (at point *G*), the time (and thus phase) varies slowly, and constructive interference is possible. From Feynman, 1988.

distance for paths right near this minimal configuration is also minimal. Just like a rounded hill is flattest near the top, and a rounded valley is flattest near the bottom, the phase changes the least when the angle of incidence is near what we call the angle of reflection. Because the phases are similar, the light waves add constructively and it appears that light is only reflected when the angles are equal.

This may all sound like shooting a sparrow with a cannon, but it is a better way of thinking about reflection for a couple of reasons. First, it sidesteps the belief that there are many separate laws about how light behaves. Second, it allows us to better understand the structural colors of animals, which we get to later in this chapter. Finally, it predicts what we see in less common circumstances. For example, suppose we had a really small mirror, only a hundredth of a micron across. The usual rule would say that we would only see reflected light if we put it in the right place (where angle of incidence equals angle of reflection), but in fact we see reflected light no matter where we place it. This is because, with such a small mirror, the path lengths are nearvly all the same regardless of where we place the mirror. So the phases add up and we see a reflection. Make a mirror small enough and it doesn't matter where we put it.

It also better explains reflection from nonmetallic smooth surfaces like glass and water. We usually say that the light bounces off these surfaces, but why should a light wave bounce off the surface of a transparent object? It doesn't. Instead, the light travels into the water or glass and is scattered at every location. It's messy to prove, but all the scattered waves cancel out except for two: (1) the reflected light where the angle of incidence and reflection are equal, and (2) the refracted light, which we discuss later in the chapter (Doyle, 1985).

Despite the fact that animals do not assemble metals into structures, they are remarkably good at making mirrors. However, they are seldom called mirrors. Instead, we hear about tapeta in eyes, silvery sides in fish, reflecting scales in butterflies, iridophores in cephalopods, and structural colors in a host of animals. So, before we get into details, we should think about why these mirrors are even made. After all, as we discussed in chapter 4, any thick, complex, nonabsorbing substance (like connective tissue) reflects most of the light that hits it. Reflected light from the structures listed above, however, is often different from this sort of diffuse reflectance in three ways. First, the reflectance has at least some directionality to it. While few animal mirrors are so good that you can see your face in them, they usually reflect most of their light in one direction. This is what gives them that shiny look. Second, animal mirrors are more efficient reflectors than bulk tissue. In other words, they can be thin and still reflect most of the light. Finally, due to wavelength-dependent interference, reflected light from animal mirrors can be colored for reasons that have nothing to do with absorption. The reflected color can also be highly saturated because the wavelength selectivity of interference can be greater than that of the absorption bands of typical biological molecules.

Animal mirrors have been studied for a long time with a recent explosive renaissance (likely driven by advances in computing power that allow analysis of more complex structures), so I can only give the briefest introduction to their diversity and function. Entire books have been written on butterfly iridescence alone.

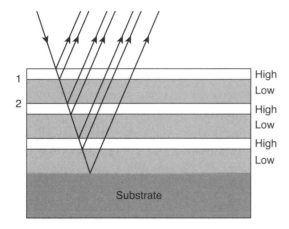

Figure 6.6: Light interacting with a quarter-wave
stack.

Structurally, animal mirrors can be divided into two categories: (1) Bragg
stacks, and (2) everything else. A Bragg, or quarter-wave, stack is an alternating pile of high and low index layers (figure 6.6). Each of these layers has a
thickness such that the light of one particular wavelength (let's say 500 nm
measured in air) only goes through a quarter of its cycle as it passes through
the layer. Remember that, because wavelength is affected by the index of the
medium, the physical thicknesses of the high index layers will be less than
the thickness of the low index layers. Two "rules" you need to know are that:
(1) the phase of light changes by 180° when it reflects from an interface
where the index goes from low to high, and (2) the phase of light does not
change when it reflects from an interface where the index goes from high to
low. Given this, let's assume that the incident beam arrives at the top surface
of the stack with a phase of zero. It is going from low to high, so the phase of
the reflected beam from the top surface (measured just above this surface) is
180°. Most of the light will not be reflected from this uppermost surface,
though. The rest goes farther in, and some is reflected at the next interface
(interface 1). Now the interface goes from high to low, so there is no phase
change, but the reflected light has traveled through the top layer twice, so
when it exits the uppermost surface, its phase is also 180°. Again, though,
only some of the light was reflected at interface 1. The rest continues down
and some of it is reflected at interface 2. We are going from low to high again,
so the phase changes by 180°. The total path to get that far in and back out is
one whole wavelength, so the phase of the beam leaving the uppermost surface is again 180°. You can keep going for as many layers as you want and will

find that the phase of the light leaving the uppermost surface is always 180°. Therefore, all the reflected beams are in phase and reflection is high for that one wavelength.

This classic explanation sweeps two things under the rug. First of all, it ignores the fact that light is also reflected by the interfaces as it is traveling back up. The whole thing still works as a wavelength-specific reflector, but to calculate the reflectance accurately you have to take these extra reflections into account. Doing it right for a stack of more than a couple interfaces involves what is known as transfer matrix theory, which isn't as intimidating as it may sound. Second, the phase change "rules" are arbitrary and make it appear as though only the interfaces matter. In reality, the entire object is scattering light, but the phase relationships between the scattered waves are such that you can pretend only the interfaces matter as long as you include these two phase rules.

Regardless of how you choose to analyze it, though, quarter-wave stacks are highly efficient at reflecting a range of wavelengths. As I said, the exact reflectance spectrum for a given stack takes some effort to determine, but there are a few trends. First, adding more layers to the stack increases the total reflectance. Increasing the difference of the two refractive indices also increases the total reflectance, but it actually lowers the saturation of the reflected color. In other words, for a stack with a given number of layers, playing with the refractive indices of the components is a trade-off between the brightness and the saturation of the reflected light. If you want only a narrow range of wavelengths to be reflected, the stack needs to be made of materials with fairly similar indices, which means that the reflectance will be dim unless the stack is thick. This is partially why highly saturated structural colors from, for example, butterfly wings are often quite dark.

Despite what people often say about Bragg reflectors, they do not have to reflect only a narrow range of wavelengths. If the differences in the two indices are large enough, you can make a Bragg stack that looks white. For example, a quarter-wave stack for 550 nm light made out of thin sheets of glass ($n \cong 1.5$) alternating with air will strongly reflect light from 440 nm to 660 nm and thus look white. Differences this great are hard to achieve in tissue, though (the guanine in shiny fish scales comes close), thus most "white" biological mirrors are deep stacks with many different spacings between layers, so that all wavelengths are reflected well.

A final issue with Bragg stacks is that the reflection is angle dependent. This occurs because the oblique path through each layer is longer than the perpendicular one and therefore a quarter of a wavelength for a longer wavelength of light. Therefore, one way to tell whether you are dealing with a classic Bragg stack is to tilt the surface. If the reflected light changes color, then you are likely dealing with a simple quarter-wave reflector. If it does not change color, a more complex geometry is at work.

Moving from Bragg stacks to "everything else," the first thing to realize is that, regardless of the complexity of the structure, all of the more complex mirrors work by the same fundamental process. Light enters the tissue (which usually has low absorption) and is scattered from every location. Depending on the regularity of the structure, there may or may not be a simple way to estimate the reflectance, but the physics itself is unchanged. This seems obvious, but is nearly always overlooked. Instead you hear people argue about whether something is a Bragg reflector or a coherent scatterer or a diffractor, and so forth—a classic case of confusing a phenomenon with the methods used to study it.

This said, analyzing the more complex structures can be a bear. Nature seems to be all about variations on a theme and hierarchical structure, so people analyzing structural colors face a bewildering variety of ultrastructures (Plate 4) (reviewed by Kinoshita et al., 2008). In some cases you find multiple Bragg stacks jumbled together at various angles (much like stacks of paper in the Dumpster behind a stationery store). In others, you find stacks with variable spacings or curved layers. Then there are tissues that don't have layers at all. Instead there may be a spongelike network of two substances, such as is found in the colored feathers of some birds (Prum et al. 1998, 1999) or high index structures shaped like spindles, ovals, ellipses, trefoils, and a host of other shapes embedded in a low index medium (e.g., Holt et al., in press). It also appears that some of these structures are dynamic, with the animal able to control the spacing, shape, and density (and thus the index) of the material at will (e.g., Izumi et al., 2010). Open a recent issue of *Nature* or *Science* and there's a good chance you will see evidence of a new optical structure in some animal.

Whether these more complex structures have specific optical advantages is a new and exciting topic of research. The spongelike networks in the bird feathers have the advantage that they reflect roughly the same spectrum, regardless of the angle of the light and viewer, unlike simple Bragg reflectors, which are highly angle dependent. This is an obvious advantage if you want to send a constant signal regardless of the location of your viewer, though you often pay a cost of reduced saturation for a given thickness of structure. Certain structures found surrounding the eyes of squid appear to be especially good at reflecting all wavelengths equally (Holt et al., in press). Other structures in stomatopods have interesting effects on the polarization of the reflected light (Chiou et al., 2008). It is likely that other optical functions will be found, though it is certainly possible that some complex structures are simply variations on the Bragg stack theme.

The ecological functions of biological mirrors are as varied as their structures and locations. Perhaps the most common function—proven or hypothesized—is signaling. These signals are usually relatively simple Bragg stacks that reflect a narrow range of wavelengths, typically at the lower end of the visible spectrum. For obscure reasons, long-wavelength colors in animals (yellows, oranges, and reds) are typically based on absorption, and short-wavelength colors (blues and purples) are typically based on scattering. Blue colors based on absorption certainly exist, as do red colors based on scattering, but for some odd reason they are both rare. Regardless, though, "structural colors" (as "scattering/interference-based reflection" is called in this context) are impressive signals, due to their high reflectance and saturation. Structural colors are found in a diverse array of species, especially in insects (see Suichi Kinoshita's beautiful book *Structural Colors in the Realm of Nature* for a review). While an actual signaling function has only been proven in a few species (as part of the extensive literature on colored signals in animals), and structural colors may be incidental in certain species, it is hard to look at the blazing metallic green of the highly aggressive tiger beetle (*Cicindela* sp.) and not see a signal. Even better than the tiger beetle, though, is the Panamanian tortoise beetle (*Charidotella egregia*), which—as Alison Sweeney put it—has the color and sheen of Harry Potter's golden snitch. If you touch this gorgeous animal, though, it goes from metallic gold to flat red in under two minutes, possibly by changing the hydration of its cuticle, which then disrupts the Bragg reflector responsible for the golden color, revealing an underlying red pigment (Vigneron et al., 2007) (Plate 5). The difference between the gold and red state is so great that for years this animal was thought to be two different species. Why it changes color after being touched is anybody's guess, but I would be shocked if it

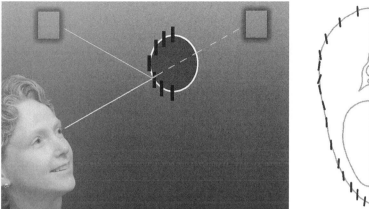

Figure 6.7: (Left) The scene reflected from vertical mirrors under water can look just like the scene behind the animal because the underwater light field is often nearly symmetric around the vertical axis. (Right) Cross section of the fish *Alburnus alburnus* showing the orientation of the reflecting plates in the scales. After Denton, 1970.

didn't have a signaling function. In general, beetles are a fertile ground for research on the ecological functions of structural color, combining diverse optics, species, and ecological niches all in one group (see Seago et al., 2008, for a recent review).

Broad-band biological mirrors are seldom used for signaling, perhaps because the one great advantage of saturated structural colors is their invariance under different optical environments (a blue butterfly wing looks blue under almost any natural light). Instead, broad-band reflectors seem to be used for camouflage, especially in pelagic habitats. At first glance, mirrors are not an obvious camouflage tactic (sardines are certainly easy to find on a plate). They work, however, because the underwater light field is more or less symmetric around the vertical axis (figure 6.7). Therefore, if you look at a vertical mirror, the reflected radiance you see nearly matches the radiance you would see if you were looking straight through the animal. However, this is only successful if the mirror is vertical. Careful research by Eric Denton and Mike Land has shown that the reflecting structures in fish are indeed vertical, which is a good trick because many fish have curved bodies. They do this by orienting the reflecting structures at different angles relative to the scales to compensate for the curvature (Denton, 1970; Denton and Land, 1971).

In addition to the silvery sides of some fish, camouflaging mirrors are found on the opaque guts of some transparent species (Seapy and Young, 1986), and surrounding the eyes of many fish and cephalopods. They are rarer on land, but it has been suggested that the metallic golden chrysalises of many butterflies, while conspicuous in the lab, are cryptic in nature because they reflect the leaves around them. This golden color is fairly common; in fact the word "chrysalis" is derived from the Greek word for gold.

Reflective structures are also found behind the retinas of many nocturnal and crepuscular animals. These are called "tapeta" and are responsible for what is often called "eye glow" or "eye shine." Anyone who has looked at a cat in a dimly lit, but not completely dark, room is well aware of these mirrors. Because they reflect incoming light back through the retina, they double the length of the light path through the photoreceptors and increase absorption. Seems simple and clever, but there are a number of associated costs. First, any light not absorbed by the retina on the second pass will bounce around the eye. Much of this light leaves the eye and makes the animal more conspicuous. We used to find our wayward cat at night by waving a flashlight until it caught her eyes. In fact, tapeta are so common on land that you have to wonder why bioluminescent searchlights never evolved in nocturnal, terrestrial predators as they did in the deep sea. Some of the reflected light never leaves the eye, but is instead absorbed by a second photoreceptor far from the original spot where the incoming beam first hit the retina. This degrades image quality. D'Angelo et al. (2008) suggested that these two problems explain why tapeta are seldom broadband reflectors, but instead reflect narrow wavelength bands that are close to the absorption peaks of the photoreceptors in the retina. This gives the reflected light the best chance of being absorbed by the original photoreceptor and thus the least chance of continuing on to cause mischief.

Something else once occurred to me as I was walking through the Woods Hole Public Aquarium. There was an orange fish that had an orange tapetum (no I don't remember the species; it haunts me). From certain angles, the tapetum did not reflect much light and you could see the pupil. But from many angles, the orange reflection completely matched the skin of the fish and made the eye disappear. It was a remarkable trick and possibly quite useful, since eyes are conspicuous and can give away the presence of an otherwise camouflaged animal.

Some tapeta can also apparently change with the seasons. In 2003, Dukes et al. presented a paper at a neuroscience conference showing that reindeer tapeta are yellow during the summer and blue during the winter. These animals live near the Arctic Circle, so summer and winter are quite different. As we discussed in chapter 3, both moonlight and starlight are redder than daylight. However, as we also mentioned, twilight is intensely blue and long at polar latitudes. In fact, near the Arctic Circle, where the sun is just barely below the horizon for the entire winter solstice day, twilight can be the dominating light environment for much of the winter. So it appears that these animals are responding to the changing light over the year to maximize their sensitivity. A fascinating master's thesis by Sandra Siefken at the University of Tromsø in Norway (the natural place to study reindeer) showed that the changing colors of the tapeta are controlled by an endogenous rhythm and appear to be mediated by a change in collagen spacing in the mirror (Siefken, 2010).

The true tapetal honors, though, go to those found in the eyes of scallops. Unlike clams, oysters, and mussels (and for reasons that escape me), Americans don't eat entire scallops, but only the adductor muscle. So we never see that scallops have fifty to one hundred eyes that peek out through the gape between the two shells (figure 6.8). The eyes vary in size and color, but in some species, such as *Argopecten irradians*, they are about one millimeter across and a stunning blue. Each eye has a lens, but early work by Mike Land, a pioneer in the field of comparative visual optics, showed that the lens is not powerful enough to focus light on the retinas (oddly there are two retinas, as different as can be in all respects). Instead the tapetum behind the retinas is of such high optical quality that it focuses the light onto the retinas. The lens only appears to compensate for the fact that the mirror is spherical and not a parabola, as a perfect focusing mirror should be. This is one of only two cases in the animal kingdom where an animal has made a mirror of optical quality. Why the eye needs two retinas, why it uses a mirror, and why what is essentially a glorified clam needs fifty to one hundred good eyes are open questions that another of my former students, Dan Speiser, took on. My favorite experiment of his involved showing scallops movies of food (in the form of particles moving on a computer screen). The scallops, held in little seats, would open their shells to feed if the particles were big enough and not moving too fast, suggesting that at least one function of their eyes may be to as-

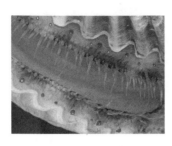 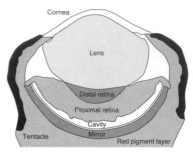

Figure 6.8: (Left) The blue mirror eyes of the bay scallop *Argopecten irradians*. (Right) Cross section of the eye from a related scallop species, the sea scallop *Placopecten magellanicus*, showing the two retinas, lens, and mirror. Right image courtesy of Dan Speiser.

sess conditions for filter feeding (Speiser and Johnsen, 2008). It was a classic case of an experiment that nobody (including me) ever thought would work, but did anyway.

The other optical-quality biological mirror is found in the deep-sea spookfish *Dolichopteryx longipes*. In addition to two relatively normal upward-looking eyes, the spookfish has two downward-facing eyes that have no lens at all (figure 6.9). Instead, the eyes have mirrors that not only focus the light, but turn it 90° so that it reaches two sideways-facing retinas. In this way, they solve one of the classic problems in reflective optics, how to keep your detector from blocking the incoming light. You have probably noticed that most large telescopes are based on mirrors instead of lenses. There are good reasons for this. Mirrors don't have chromatic aberration (they focus all wavelengths to the same point) and their location at the bottom of a telescope makes them easier to support. However, they focus light directly back into the beam path. This means that you have to mount a detector or a mirror in the middle of this path to either collect the light or bounce it out of the path to a detector. Not only is this awkward, but it inevitably blocks some of the incoming light. The scallop eyes we mentioned above have a bad case of this, where light has to travel through both retinas unfocused before it can be focused by the mirror and come back. This almost definitely degrades the image and lowers contrast. One way humans get around this problem is to use an off-axis mirror that has an asymmetry to it so that reflected light goes slightly off to the side. These are tricky to make and expensive. As usual, animals have beaten us to it.

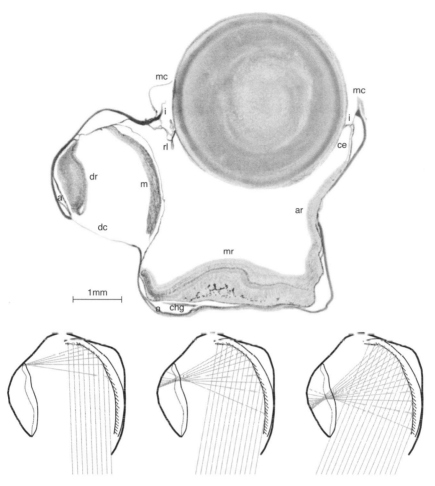

Figure 6.9: (Top) Cross section of the eye of the spookfish *Dolichopteryx lon-gipes*. The right part is a typical camera eye with a lens. The left part is a mirror eye. (Bottom) Incident light from three different angles of incidence being focused on the retina by the off-axis mirror. From Wagner et al., 2009.

DOES LIGHT BEND? REFRACTION

Let's leave mirrors now and go back to our night dive. Suppose your dive buddy surfaces and shines the flashlight into the water from a boat. To you, it looks like the beam bends when it enters the water (you don't actually see the bend; instead it looks like the beam is coming from a different spot than where your buddy's light should be). The angle over which the light bends depends on the ratio of the refractive index of the water to that of the air m and is given by Snel's law (yes, this is how it's really spelled):

$$\theta_{refracted} = \sin^{-1}\left(\frac{\sin\theta_{incident}}{m}\right). \qquad 6.2$$

But why does light bend in this way? You may have heard of Fermat's Principle of Least Time, which says that light chooses a path that takes the least time. It turns out, if you take into account that the phase velocity is lower in water than in air (due to the higher refractive index), the path the light follows does indeed take the least amount of time. But why is this important, and is light truly doing a set of trigonometry problems before choosing where to go? As with the mirror, light goes everywhere between the flashlight and your eye (figure 6.10). However, the paths that are close to the one given by Snel's Law have nearly the same travel time, and thus nearly the same phase. So they constructively interfere and it looks like the light bends at the surface of the water. The fact that it's the minimum time doesn't matter. Again, what matters is that the *change* in time for paths near this minimum is also minimal.

There are at least two interesting examples of refraction in biology. The first is Snel's Window. Because the light is refracted at the air-water interface, the entire hemisphere of the sky (as viewed by someone under water and looking up) is compressed into a cone about 96°across. If you look at angles farther from the zenith than that, you can't see what's above the water. Instead you see a mirrorlike internal reflection. This refraction of the entire hemisphere into a smaller field of view implies that the sun is never that far from directly overhead, even near sunset. It also means that you have this funny view of a whole hemisphere of sky and landscape that you can take in at once. The average human cannot see a whole hemisphere at once, at least not clearly, but we can see all of Snel's Window. This could have important implications for underwater animals using celestial navigation, but, before you get too excited, it is important to realize that Snel's Window is seldom as distinct as the books make it out to be. Nearly every time I am under water, I look for Snel's Window and have only seen excellent cases of it a few times out of hundreds. You need clear water with few to no waves. The sun also has to be low in the sky, or looking up blinds you. Finally, you need some nearby

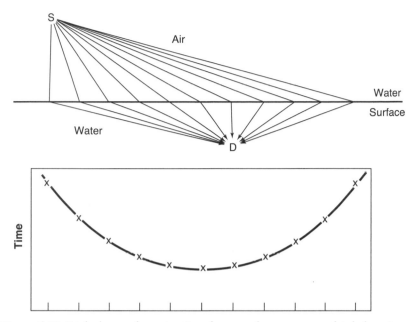

Figure 6.10. Light leaves the point S and enters the water at multiple locations. For the path determined by Snel's Law, the time it takes the light to travel from S to D is least and the change in time for neighboring paths is close to zero, leading to constructive interference and the appearance that light bends at the interface. From Feynman, 1988.

structures to see, like trees or a dive boat. So the best places are clear, quiet ponds surrounded by trees. Interestingly, the animal most famous for dealing with refraction at the air-water interface, the archer fish, lives in exactly this sort of habitat. It squirts water at bugs hanging from low branches and is able to compensate for refraction. This is clever, but I doubt it would be possible in a more typical air-water surface that is ruffled by waves.

The second important case of refraction in biology is found in lenses. In the standard explanation of lenses, light bends as it enters the lens and then bends again as it leaves it. If all the light comes from an object fairly far away, then all the bent rays from it meet at a spot called the "focal point," where the image is in focus. As we discussed, though, ray tracing (where you follow thin beams of light using geometric optics) is just a useful mathematical shorthand and metaphor for what is actually a case of scattering and interference within the lens. Let's assume that the lens has a higher refractive index

than the surrounding medium. So the phase velocity is lower and the wavelengths are shorter inside it. If you want to have all the light from one point
reach another point at the same time (and thus be in phase and constructively interfere), you have to put a lot of glass between the two points for the
shorter straight path and progressively less glass for the longer paths that are
off the center line (figure 6.11). Do it just right and you have a lens. In other
words, a lens is not bending individual rays of light so much as it is making
sure that all the in-phase light from one point makes it to another point still
in phase. To prove that none of this has anything to do with lenses bending
light, you can replace the lens with a mask. This mask, which looks like a
concentric set of rings, only lets through light that has path lengths different
from one another by one full wavelength of light. So even though all the
paths don't have the same length, they end up having the same phase and
thus add constructively (figure 6.11). This little gadget is called a "Fresnel
zone plate" and actually works (I have made and used them). They are used
in X-ray microscopes and in the telecom industry, but, as far as I know, there
are no biological examples of these, which is a shame because there is no flat-
ter way to make a lens, even though it only works well for one wavelength
and has multiple focal lengths.

Now, you might be saying, "But a zone plate works by diffraction!" Again,
diffraction is just another case of scattering combined with interference. But
what is doing the scattering in this case? What you will often see in textbooks is Huygen's Principle, which states that each gap in the zone plate (or
any hole in an opaque wall) is the source of a spherically expanding wave.
This explanation works mathematically, but makes no physical sense (at least
to me). A hole cannot be the source of light, scattered or otherwise. It is just
a hole. Bohren and Clothiaux (2006) explain it this way: Imagine you shine
a beam of light at an opaque black sheet of paper. You see no light on the
other side. One way to think of this is that light is composed of photons that
were absorbed by the sheet of paper. Another way to think of it is that the
light wave also exists on the other side of the sheet, but is completely can-
celed by a wave scattered by the sheet itself, leading to total destructive interference. This might sound insane and—believe me—I wrote Craig Bohren
many letters where we argued this. I still have my concerns, but remember
that absorption and scattering are intimately related and that the interference of the incident waves and the scattered wave in glass is what makes it
look like light is traveling more slowly in the material. So it is not *too* big a

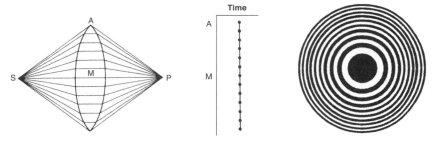

Figure 6.11: (Left) All the paths from S to P take the same amount of time, so there is strong constructive interference at P and the light passing through the lens is focused. From Feynman, 1988. (Right) Zone plate. The pathlength of light passing through each clear ring differs from that of light passing through the neighboring clear rings by exactly one wavelength, so the object behaves like a lens.

stretch to imagine that opaque substances emit waves that cancel the incident waves. If you do accept this explanation, then diffraction makes sense, because by placing a hole in the black sheet, you have destroyed the perfect destructive interference. The way in which it is destroyed leads to the patterns you see.

Biological lenses work just like human-made ones, but come in various types. Those found in the eyes of humans and many other terrestrial vertebrates are shaped like the classic magnifying-glass lens. Those found in aquatic vertebrates and cephalopods, though, are often spheres. This is because the large refractive-index difference between air and tissue gives the corneal surface of terrestrial animals a lot of refractive power. In fact over three-quarters of the refractive power of our eyes comes from our cornea, which is why the slight alterations achieved in LASIK and other corneal surgeries are so effective at fixing vision problems. However, the difference between the refractive indices of water and tissue are far less, so the corneas in aquatic animals function primarily as windows, leaving the lens to take on the primary job of refracting the light. The only way to do this without having a truly gigantic eye is to make the lens a sphere. However, one glance through a clear marble shows that a sphere is a terrible lens. Light passing through different parts of it are focused to entirely different places, a phenomenon known as "spherical aberration." Aquatic animals solve this problem by smoothly changing the refractive index of their lens from a high value in the center to a low one near the periphery. The gradient can actually be quite steep, going all the way from 1.55 to just above 1.33, and has to be of just the right shape to work. Calculating the path through a graded-index lens is not as simple as calculating the path through a lens with one index and definitely departs from the idea that light is just bent at the surfaces. Remarkable as they are, graded-index materials appear to have evolved a number of times and are not found only in spherical lenses. They are also seen in the cylindrical lenses of the compound eyes of certain insects (e.g., scarab beetles [Caveney and McIntyre, 1981]) and in the tapered fiber optic cables of the *Phronima* eyes described above (Land and Nilsson, 2002).

Due to "diffraction," small lenses, particularly those in the compound eyes of insects and crustaceans, run into trouble. They are so small that their ability to focus light is greatly limited by this spreading. So any animal with a compound eye that wants sharp vision needs to make the lenses larger, which quickly leads to enormous eyes. Compound eyes that could see as sharply as our actual human camera-like eyes would have to be one meter across (Land and Nilsson, 2002). This calculation accounts for the lower resolution of the nonfoveal parts of our retina; a compound eye with the resolution of the fovea in every direction would be twenty meters across! All this means that how biologists look for foveas in compound eyes is different from how they look for them in "camera" eyes. In camera eyes (like ours), the fovea is where the photoreceptor density is the highest. In compound eyes, the fovea (if there is one) is where the facets are the largest.

Land and Nilsson's *Animal Eyes* does a wonderful job of reviewing the lenses and optics of a diversity of species, and there is no point in reproducing it here. Instead, let's move on to photonic crystals.

PHOTONIC CRYSTALS

Scientists are no strangers to fashion, and, over the past fifteen years, biology has been going through a word revolution, replacing "ologies" for "omics." "Molecular biology" has been largely replaced with "genomics" and "proteomics," and "physiology" is fighting for its life against "metabolomics." I have suggested replacing "organismal biology" with "orgonomics" and "ecology" with "economics," but these two words seem to be taken. Physicists have taken a similar path, swapping out "optics" for "photonics." New words are fun, but it is important to realize that in most cases, these changes, where they have any meaning at all, are more about advances in methods than revolutions in knowledge.

Which brings us to photonic crystals. Probably no subject in modern biological optics is described with more mysticism and less clarity than photonic crystals. Much of this stems from the fact that the most useful math to analyze photonic crystals is unfamiliar to biologists. However, I feel that a lot of the confusion is unnecessary.

First of all, a photonic crystal is just another repetitive structure where you have to take interference into account to determine the observable scattering. Second, one-dimensional photonic crystals are familiar to just about everyone; they are soap films, Bragg stacks, and other interference reflectors. These structures, as we now know, strongly reflect certain wavelengths of light and let others pass through via interference. They are called one-dimensional because the index only varies along one axis, even if the material is formed in sheets. However, this is where terminology once again steps in and kills the fun. The wavelength region where light is reflected is now called the "photonic band gap" and you begin to read sentences like "The photonic band gap (PBG) is essentially the gap between the air-line and the dielectric-line in the dispersion relation of the PBG system" (from Wikipedia). Not helpful.

This does not mean that photonic crystals themselves are not interesting. They are—especially the higher-dimensional ones. As we discussed, the problem with a Bragg stack is that the light it reflects depends on the orientation of the incident beam relative to the layers. Turn a Bragg stack ninety degrees and it doesn't work at all. However, two-dimensional photonic crystals have the same optical properties (typically highly selective transmission or reflection of light over narrow wavebands) for light with any incident direction within one plane. Three-dimensional crystals work for light coming from any direction at all. This is a good trick and is achieved via repeating structures of high and low refractive index. Opals are natural, though imperfect, three-dimensional photonic crystals, consisting of a myriad of closely packed spheres of silica, each about 150 nm to 300 nm in diameter. The light scattered from these spheres creates a complicated interference pattern that is strongly wavelength dependent, and gives opals their color.

Optics engineers hold photonic crystals in awe, because it is difficult to make the required structures on the tiny size scales needed to work with visible light. The tiny holes and patterns needed are most often made using techniques borrowed from the semiconductor industry, but it is still a technological hurdle.

However, organisms have no trouble making periodic structures of the required size scales—it is one of the things nature does best. So, since photonic crystals were first named in 1987, they have been found in a number of animal tissues. Of course, one-dimensional photonic crystals are everywhere in animals, but higher-dimensional ones have also been found, one of the first in the spines of the polychaete *Aphrodita*. This stubby brown blob, commonly known as a "sea mouse" (for reasons that are clear if you look at it and know your Greek mythology and sailors slang), is about the last place you would expect to find a beautiful optical structure. However, its spines are highly iridescent when viewed under bright directed light. Andrew Parker and his colleagues looked at the ultrastructure of the spines and found that they operated as two-dimensional optical crystals, with 100% reflectance at some wavelengths and next to no reflectance at others (Plate 6) (Parker et al., 2001). They suggest that the iridescence may function in species recognition or courtship, but I have my doubts. These animals are usually found buried in muddy bottoms, sometimes at great depths. Once you get more than about five meters below the surface in even the clearest of waters, the light field is simply too diffuse and monochromatic to create any iridescence in even the most optically exciting of structures. My guess is that the photonic crystal is simply a pretty byproduct of a structure that gives the animal a strong spine, much like the pearly interiors of abalones. The iridescence from the comb rows of ctenophores, another photonic crystal, also likely has no optical function since it can only be seen near the surface (also the animal itself is blind and has no toxicity to warn others of).

However, there are photonic crystals with obvious optical functions. They are primarily found in butterflies and can be quite complex, combining scattered light with other reflections and with absorption by pigments. These are also three-dimensional, which is rarer than the two-dimensional and of course the one-dimensional forms. In addition to ctenophores and *Aphrodita*, two-dimensional crystals are also found in the feathers of peacocks and some other birds. The crystals in birds and insects seem to function as structural colors for conspecific signaling, though of course we don't have definitive behavioral evidence for every species. It is important to realize, though, that the light from photonic crystals is not special, and so colors based on them likely serve the same functions as more typical structural colors. In other words, the animals don't "know" that they have evolved a structure that makes an optical engineer's mouth water.

Conclusion

In the end, nearly all the optics you learned in high school and college is the result of scattering and interference. So why wasn't it described this way then? After all, physicists are usually all about unity. I suppose there could be a number of reasons, but my hunch is that it's because these optical principles and shortcuts were developed before computers. Many of the scientists that developed these rules understood the underlying physics, but had no way of actually doing the calculations. So they developed equations that were either approximations, or worked only in specific geometries. These people were geniuses at getting maximum information with a minimum amount of calculation, but the result is that many people are left with the idea that reflection, transmission, refraction, diffraction, and a host of other phenomena are separate processes, when in fact they can all be explained via scattering and interference.

So why not just use these simple rules? In many cases you can and should. Unfortunately, unlike materials in the technological world, biological tissue is a geometric mess. Optically perfect spheres, parabolas, and flat surfaces seldom exist in nature. More importantly, tissue is seldom homogeneous. This means that applying classical optical equations such as Snel's or Fresnel's is perilous at best, pointless at worst. Instead, you often do need to consider how much light is scattered from every little region and add up all the waves. Sounds horrifying, but few people actually do this analytically. Instead they use software packages that do this via standard finite element and finite difference methods. Describing these methods is beyond the scope of this book, but recent packages are not too expensive and allow you to input any geometry you want. The boom in personal computing has now given us cheap machines that can grind through these numerical algorithms in minutes rather than weeks. In other words, the analysis of complicated optical structures has now become feasible and nearly routine in some cases. This is fortunate, because we have only just begun to see what animals can do with light.

Further Reading

Feynman's *QED* (described at the end of chapter 1) does a great job of unifying geometrical optics under a simplified version of quantum mechanics. Short and inspirational.

A more classical explanation that is similarly inspiring and thought-provoking can be found in Bohren and Clothiaux's *Fundamentals of Atmospheric Radiation* (also described at the end of chapter 1).

Suichi Kinoshita's *Structural Colors in the Realm of Nature* is beautiful and comprehensive on the topic.

If you read French, Serge Berthier's *Les Couleurs des Papillons* is a thorough analysis of structural colors in butterflies. Even if you don't read French, you can get a lot from the beautiful figures.

Land and Nilsson's *Animal Eyes*, described before, is the place to go for the optics of animal eyes and lenses.

Plate 1: The aurora borcalis over Bear Lake, Eielson Air Force Base, Alaska.
Photo courtesy of the U.S. Air Force, taken by Senior Airman Joshua Strang.

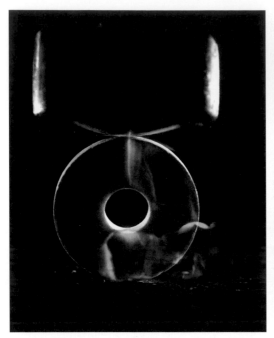

Plate 2: (Left) Triboluminescence from a Wint-O-Green LifeSaver being struck by a hammer. Courtesy of Ted Kinsman. (Right) Triboluminescence from Scotch Tape peeled in a vacuum. From Camara et al., 2008.

Plate 3 *(opposite)*: Assemblage of transparent animals. (A) *Amphogona apicata* (hydromedusa), (B) *Amphitretus pelagicus* (pelagic octopus), (C) *Eurhamphaea vexilligera* (oceanic ctenophore), (D) *Planctosphaera pelagica* (hemichordate larva), (E) *Naiades cantrainii* (polychaete), (F) *Phylliroë bucephala* (pelagic nudibranch), (G) *Pterosagitta draco* (chaetognath), (H) *Carinaria* sp. (heteropod), (I) *Bathochordeus charon* (larvacean), (J) *Periclimenes holthuisi* (anemone shrimp), (K) *Bathophilus* sp. (larval form of deep-sea fish), (L) *Cardiopoda richardi* (heteropod). Images courtesy of Laurence Madin, Steve Haddock, Jeff Jeffords, and Edith Widder.

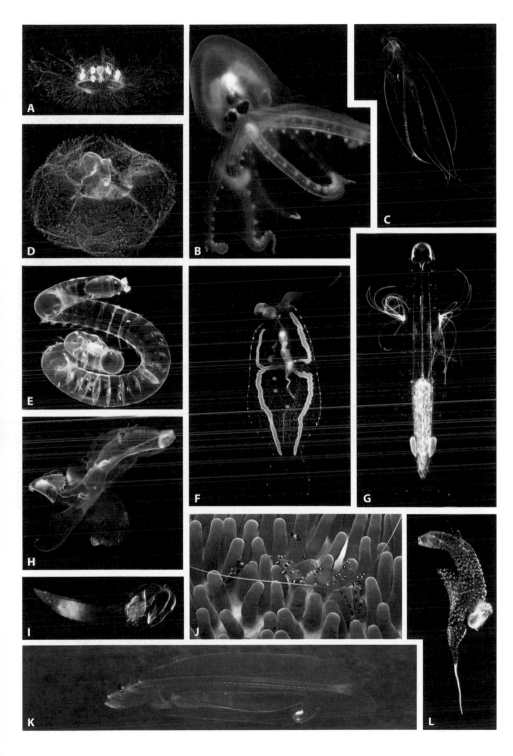

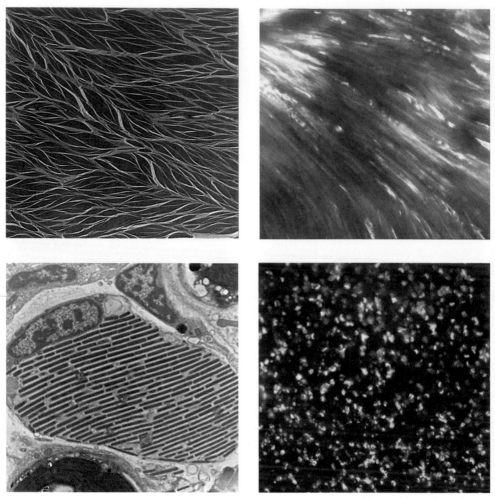

Plate 4: The left column shows electron micrographs of various photonic structures in cephalopods, and the right column shows photographs of the resulting optical effects. Courtesy of Alison Sweeney.

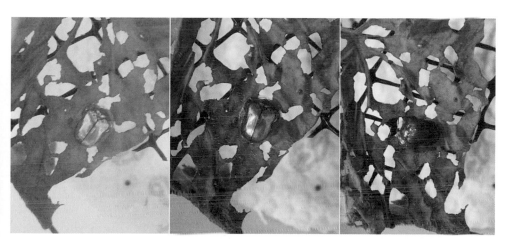

Plate 5: The Panamanian Tortoise Beetle *Charidotella egregia* changing from its typical metallic golden color to a diffuse red color after being touched. From Vigneron et al., 2007.

Plate 6: (Left) The iridescent spines of the polychaete *Aphrodita sp.* Courtesy of Michael Maggs. (Right) Electron micrograph of a cross section of a spine from this animal, showing the two-dimensional photonic crystal structure that creates the iridescence. From Parker et al., 2001.

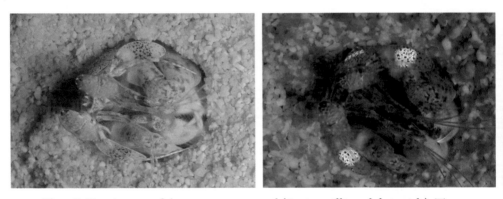

Plate 7: Two images of the same stomatopod (*Lysiosquillina glabriuscula*). The image on the left was shot under ambient light at depth. The image on the right was shot under blue light and viewed through a filter that only passed green light. The bright patches in the right picture (more muted in the left) are the yellow fluorescent spots on the antennal scales. Images courtesy of Roy Caldwell.

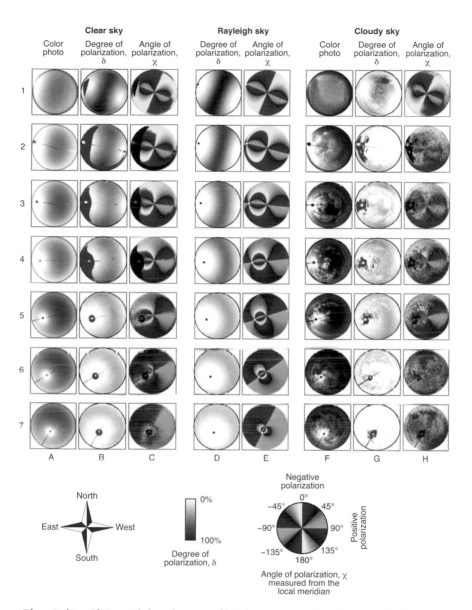

Plate 8: (A–C) Spatial distribution of brightness and color, degree of polarization, and angle of polarization in a clear sky at various elevations of the sun from dawn (1) to noon (7). (D, E) Patterns predicted for an ideal sky. (F–H) Same as (A–C), but with cloudy skies. The red or black regions of the sky in columns B, G or C, H, respectively, are overexposed. The position of the sun is indicated by a black or white dot. The radial bar in the pictures is a wire holding a small disk to screen out the sun. From Pomozi et al., 2001.

Plate 9: Circular polarization signal in the mantis shrimp *Odontodactylus cultrifer*. Bottom two images show the keel telson viewed under left-handed and right-handed circularly polarized light. [A] courtesy of Christine Huffard; [B] and [C] from Chiou et al., 2008.

CHAPTER SEVEN

Fluorescence

This isn't an office. It's Hell with fluorescent lighting.
—AUTHOR UNKNOWN

Fluorescence does not make light! There, I have been waiting six chapters to say that and finally got it out of my system. I don't have many pet peeves—I find them peevish—but this one matters. The misunderstanding comes in two flavors. The first is a common misuse of terms. Historically, bioluminescence was often referred to as phosphorescence, and many people, even biologists, still consider the terms synonymous. This is not helped by tourist brochures for local bioluminescence hotspots that refer to the places as, for example, "phosphorescent bays." Even more unfortunately, because phosphorescence is closely related to fluorescence (as we will discuss in this chapter), the latter term has joined the party and is also often used synonymously with bioluminescence. Just last night, I was at a dinner where an invertebrate zoologist of international stature referred to fluorescence as bioluminescence. Many bioluminescent animals do use fluorescence to change the wavelengths of light that are emitted, green fluorescent protein being the most famous example. Also, some luciferins are fluorescent, but the fundamental emission process is only loosely related. As we discussed in chapter 3, bioluminescence is a form of chemiluminescence in which specific chemical reactions result in the emission of photons over a certain range of energies and thus wavelengths. This is not fluorescence.

While this mislabeling is ubiquitous and, like the common cold, may never be fully stamped out, it is less pernicious than another misunderstanding, in which, while it is accepted that fluorescence is not bioluminescence, it is still believed that fluorescence makes light. Fluorescence can never make light, it can only take it away.

What do I mean by this? Since we talk about light in both energy and quantal terms, I need to be clear. Fluorescence cannot add more optical energy to the system. If you hit a fluorescent surface with an irradiance of one watt/m², the light leaving the surface can never be more than one watt/m², as this would violate the conservation of energy principle. In fact, usually less than one watt/m² will leave the surface because fluorescent substances are not 100% efficient. Some of the energy is converted to other forms. However, if you think about it in photon terms, it is possible to add more photons to the system. For this to work, the new photons must have a lower energy than the original photons. For example, crystals of beta barium borate can convert one photon into two photons, each with half the energy of the original photon. I am not sure if any biological fluorophores do this, because the emitted photons usually have energies not much lower than the incident photons, but I suppose it is theoretically possible for there to be a biological molecule that increases the number of photons. However, even this remote possibility is not the same as creating light.

Despite this, it is not uncommon for biologists to talk about fluorescence as if adding it will make something glow. My guess is that this is mostly due to the way most biologists view fluorescence, which is through a fluorescence-equipped microscope. Look at a fluorescently labeled tissue under a confocal or standard fluorescence microscope and it does indeed look like you have made light. However, what do you don't see is the incident light that excited the fluorescence. This light is absorbed by the fluorophore and whatever remains (which is usually a lot) is screened out by a highly selective filter between the sample and your eye. Take away this filter, and everything looks decidedly less impressive.

The filtering technology we now have has allowed scientists to make striking images. Images of animal fluorescence, UV coloration, tissue birefringence, or even iridescence from ctenophore comb rows are beautiful if done with the right filters and lighting. However, the natural world seldom has light with these special characteristics, so it is important, when trying to assess the ecological importance of an animal's optical properties, to not fall in love with the image and to make sure you find out what it looks like under natural light. Sounds obvious, but many, perhaps most, studies ignore this crucial aspect.

Okay, rant over. So, if fluorescent molecules don't create light, what do they do? First and foremost, fluorophores are pigments—meaning that they absorb light. However, unlike most pigments that convert all the absorbed energy into other forms, fluorophores only convert part of it. The remaining

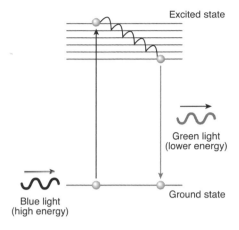

Excited state

Green light
(lower energy)

Blue light
(high energy)

Ground state

Figure 7.1: Simplified Jablonski energy diagram of fluorescence process.

energy is emitted as a photon. Usually this photon has a lower energy and longer wavelength than the incident photon, the difference in energies being known as the "Stokes shift." However, every now and then, a photon can be emitted that has a higher energy and lower wavelength than the incident photon. This is known as an "anti–Stokes shift." The extra energy needed for this is taken from vibrational states of the molecule, which actually cools it a bit. This somewhat violates "fluorescence doesn't make light," but is a minor part of biological fluorescence, so we'll ignore it. It is also possible, in very intense light, for certain molecules to absorb two photons and then, by combining the energies, emit a photon of a higher energy and lower wavelength. This is related to the principle behind two-photon microscopy, but I am unaware of any animal that does this. In at least 99% of biological cases, you are looking at the conversion of one photon into another photon with a lower energy and longer wavelength. You can think of this as inelastic scattering, since it looks like a photon bounced but lost energy in the process, but this term is usually reserved for other processes that we will discuss at the end of this chapter.

I'm sure there's a way to depict fluorescence in wave language, but it's nearly always shown in quantum form via an energy level diagram (figure 7.1). A photon is absorbed, sending the molecule to a higher energy level. Rather than jumping all the way back down to the ground state in one step (and emitting a photon of the same wavelength, which we would call a scattered photon), the molecules goes back to the ground state in at least two jumps, the last of which usually results in photon emission. Since this last

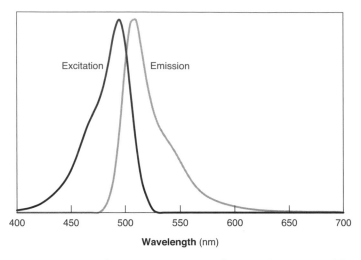

Figure 7.2: Excitation and emission spectrum of a typical commercial fluorophore. Both have been normalized to have the same peak height to better show that the spectra are nearly mirror images of each other. Data courtesy of Mikhail Matz.

energy drop is less than the original energy boost, the emitted photon has a lower energy and longer wavelength. Sounds simple, but never confuse a diagram with an understanding of what is really going on, especially in this case. Determining the energy levels of a real molecule and then figuring out which jumps are most likely to happen and where the energy goes is difficult. Predicting any of this from first principles is nearly impossible. A large and involved field of physics is devoted to these problems.

The central fact to take away from this, though, is that, while fluorescence doesn't make light, it does change how it is packaged, which has observable consequences, even without special filters that magnify the effect. Before we get into these consequences, though, we need to go over some terminology.

EXCITATION, EMISSION, AND EFFICIENCY

Many things fluoresce weakly, so when biologists talk about fluorescence, they are usually talking about the fluorescence of a specific molecule that converts a large fraction of the absorbed light into light of other wavelengths. A molecule of this sort is often called a "fluorophore." Regardless of how much light it converts, though, any molecule's fluorescence is described by three properties. The first is its excitation spectrum (figure 7.2). Though it's given a different name, this is essentially the absorption spectrum of the molecule. This makes sense, since, in order for a photon to excite fluorescence, it needs to be absorbed first. There are wavelengths of light that are absorbed but do not contribute to fluorescence (for example, those with very high or low energies), but in the near-UV/visible range for biological molecules, the absorption and excitation spectra are nearly identical. As with visual pigments, once a photon is absorbed, its energy doesn't usually affect how much light is emitted or at what wavelengths. Instead, as with visual pigments, the wavelength of the incident light only affects the *probability* that a photon is absorbed and may cause fluorescence. This fact, known as "Kasha's Rule," might seem a bit strange, since you would think that the energy of the absorbed photon would have a big impact on what could come back out. This rule is based on the fact that fluorescence nearly always happens during the last jump between the lowest possible excited state and the ground state. Therefore, as long as you have enough energy to get above this first excited state, the fluorescence is not going to change. This is not an absolute rule, however, more in the "of thumb" category.

The second defining property of a fluorescent molecule is its emission spectrum. You will notice that the emitted photons do not all have the same wavelength, despite the explanation of it all being due to the last jump between the lowest excited state and the ground state. This is due to additions and subtractions of vibrational energies and other bells and whistles that give you a hint of the true underlying complexity. The emission spectrum essentially gives you the probability that the emitted photon will have a certain wavelength. You will notice that the excitation and emission spectra look like slightly overlapping mirror images of each other. This is common in many fluorescent molecules, but not a universal rule.

The final property of a fluorescent molecule is the efficiency. This could be defined in two ways, energy emitted divided by energy absorbed, or photons emitted divided by photons absorbed. This efficiency is also called the "quantum yield," which tells you that it's the latter definition. I personally find this a bit odd, since it theoretically allows you to get above 100%. However, quantum efficiency is the accepted term and makes sense for biologists because they often care more about the number of photons than the total energy.

It is important to note that the denominator in fluorescence efficiency is the number of photons absorbed, not the number of incident photons. Therefore you cannot calculate it by measuring the light output and dividing it by the light input. You need to know how much of the light was actually absorbed. It also means that something can have a high fluorescence efficiency and only fluoresce weakly if it absorbs little light. This is another one of those cases where you must be careful. For example, the similar-sounding "quantum efficiency" does have the number of incident (instead of absorbed) photons in its denominator.

So, given these three properties, how do you actually calculate how much light is emitted at each wavelength? Because of the increasing number of studies on animal fluorescence and its possible ecological importance, it is useful to know how to do this. So I'll go into it in some detail. First, you need to calculate how much light is absorbed. Suppose you have a patch of fluorescent molecules on a microscope slide. The molecules have a molar volume density ρ and the patch has a thickness t. You illuminate this patch with a broad spectrum of light from above that results in an irradiance at the patch of $E(\lambda)$. Using the absorption equations from chapter 4, the number of photons absorbed at each wavelength is:

$$ N_{abs}(\lambda) = E(\lambda)\left(1 - 10^{-\rho t A(\lambda)}\right). \qquad 7.1 $$

The absorbance $A(\lambda)$ is by convention given in optical density units (i.e., using base-10 logarithms instead of natural logarithms). Like visual pigment absorbances, it is usually also divided into two parts: (1) a spectrum normalized to a peak of one, and (2) a coefficient. In this case the normalized spectrum is what is called the "excitation/absorption spectrum $V_{ex}(\lambda)$," and the coefficient is generally given as the molar absorption coefficient M (in other words $A[\lambda] = MV_{ex}[\lambda]$). This means you should give the density ρ in molar terms. You don't have to do it this way, but most fluorophores are described via their molar absorption coefficient, so it is often easiest. As with calculations of visual pigment absorption (see chapter 4), you need to take care with the units. Again, your guiding principle should be to make sure the exponent ends up being unitless. The molar absorption coefficient is usually given in L mol^{-1} cm^{-1}. If this is true, you are fine if you give the concentration of your fluorophore in molarity (moles/L), and your thickness in cm.

Equation 7.1 only gives you the amount of light absorbed at one wavelength, so you need to sum over a range of wavelengths. This sum is:

$$ N_{abs} = \int E(\lambda)\left(1 - 10^{-\rho t M V_{ex}(\lambda)}\right)d\lambda. \qquad 7.2 $$

Usually, going by 10-nm-wavelength intervals is precise enough, but if the excitation curve is especially steep, you may need to go to 5 nm intervals to accurately sample it.

Once you know the total number of photons that are absorbed it is pretty straightforward to calculate how much light is emitted at each wavelength. Essentially, it is just:

$$N_{em}(\lambda) = N_{abs}QV_{em}(\lambda), \qquad 7.3$$

where Q is the fluorescence efficiency and $V_{em}(\lambda)$ is the emission spectrum. There are only two remaining wrinkles, both fairly minor. First, remember that the fluorescence efficiency is defined as the total number of photons emitted divided by the total number absorbed. So, to be correct, the emission spectrum has to be normalized so that it has an integral of one.

The other wrinkle is that the fluorescence is emitted equally in all directions. This feature of fluorescence has ecological implications that we'll get to soon, but for now we need to think about how to calculate the actual radiance of the patch. Conveniently, the radiance of a patch that emits equally in all directions (known as a "Lambertian emitter") is well known and is simply the number of emitted photons divided by π. In this case, because half the photons are going in the other direction (i.e., out through the bottom of the patch and away from us), we also have to divide by 2. So the emitted radiance $L_{em}(\lambda)$ equals $N_{em}(\lambda)/2\pi$. By using this fact and combining equation 7.2 and 7.3, we get:

$$L_{em}(\lambda) = \frac{Q}{2\pi}V_{em}(\lambda)\int E(\lambda)\left(1 - 10^{-\rho t MV_\alpha(\lambda)}\right)d\lambda. \qquad 7.4$$

This may look a bit gruesome, but it is straightforward to do in a spreadsheet program or MatLab script. As with light absorption with visual pigments, though, we must work slowly and be careful with units.

How Fluorescence Can Make Things Appear Brighter

Even though fluorescence does not make light, it can nevertheless make things appear far brighter and/or more conspicuous. This occurs via several processes. First, fluorescence can convert light to wavelengths that are more easily detected by a given visual system. As we discussed in chapters 2 and 4, our eyes (and the eyes of all animals) are more sensitive to certain wavelengths of light than to others. For example, humans are about fifty times more sensitive to 550 nm green light than to 440 nm blue light and 670 nm red light. This is why green laser pointers appear so much brighter than red or blue ones. Therefore, a fluorescent surface that converts only 20% of 440 nm photons to 550 nm photons (and completely absorbs the remaining 80%) will appear to be about 10 times brighter (= 20% of 50) than a non-fluorescent surface that reflects 100% of the 440 nm photons. You haven't created more photons (in fact you have lost 80% of them), but things appear brighter because the new photons are far more likely to be absorbed by our photoreceptors. The effect is seldom as dramatic as described here, though, because the excitation and emission spectra are usually broad and overlap (see figure 7.2), so you are not converting photons of one wavelength all to another distant wavelength.

Second, fluorescence can be used to convert light into wavelengths that propagate more easily in a given medium. While air transmits light of different wavelengths relatively equally (at least over biologically relevant distances), water certainly does not. Therefore, for example, blue bioluminescence could be converted to green light via a fluorophore to transmit further in green coastal waters (where blue light is absorbed by chlorophyll). This again does not create light, but it does allow it to travel farther.

Perhaps more importantly, fluorescence can be used to create light of wavelengths that are rare in a given environment. Even clear ocean water absorbs long-wavelength light strongly, leaving little orange, yellow, and red light below even a few meters and little green light below a few tens of meters. The color of an object depends as much on the light illuminating it as on the actual reflectance, so red objects a few meters below the surface look black (lips and tongues at scuba depths look ghastly). However, a fluorescent substance can convert light of dominant wavelengths (for example, blue light in the ocean) to wavelengths that are rare, creating a highly conspicuous and specific signal. Viewers can accentuate this effect by using ocular filters that screen out the dominant wavelengths and thus increase the contrast even further. Also, because fluorescent light is scattered roughly equally in all directions, it can be even more conspicuous when looked at from certain angles (figure 7.3). Both on land and in the ocean, the overhead light is much brighter than the horizontal light, sometimes by a couple of orders of magni-

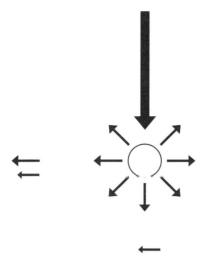

Figure 7.3: A fluorescent object in the ocean illuminated by bright down-welling light. Even though only a fraction of the light is converted to fluorescence, this light is emitted in all directions equally. Because the horizontal background light is so much dimmer than the down-welling light, the fluorescence can be brighter than the background when viewed horizontally. This effect can be magnified if the viewer is more sensitive to the wavelengths of the fluorescence than to the wavelengths of the background light.

tude. Therefore, even a relatively inefficient fluorophore can appear quite bright when viewed horizontally because it has taken a fraction of the much brighter down-welling light and converted it to light that is emitted in all directions. Even this fraction can be brighter than the horizontal background light.

Therefore, even though fluorescence cannot make light, it definitely has the potential to make conspicuous signals, especially under nearly monochromatic light environments. Bioluminescence has similar advantages, but with one big cost—it takes energy to make light. While the cost of bioluminescence is clearly not prohibitive, since it is so common in the ocean, it does appear to limit the phenomenon to environments with low light levels. Fluorescence, in contrast, can be used in the bright near-surface waters of the tropics, which have led some of us in the visual ecology community to refer to it as the poor man's bioluminescence.

Fluorescence in Animals

Even though fluorescence is not that difficult to study, we know surprisingly little about its ecology. This is odd when you consider that polarization vision and polarized signals have been fairly well studied, despite involving far hairier physics and methods. In contrast, only a handful of people have looked at fluorescent signals in a comparative and ecological context. Vastly more research in biological fluorescence has concerned itself with measuring chlorophyll in the ocean via its weak fluorescence, and mining nature for useful fluorescent labels for microscopy. This is a shame, because there is likely a lot of low-hanging fruit out there, just waiting to be plucked by researchers fitted with only light sources, filters, and clear heads.

Among animals, significant human-visible fluorescence is seen in cnidarians (corals, anemones, and hydrozoan medusae, but not scyphozoans or cubozoans), bioluminescent ctenophores, pontellid and aetideid copepods, various malacostracan crustaceans, jumping spiders and scorpions, certain polychaetes, nudibranchs, cephalopods, and various fish and birds. This list is far from exhaustive, but should give an idea of the diverse and patchy nature of the phenomenon. Before discussing possible functions, however, it is important to mention that many ubiquitous biological molecules are fluorescent (figure 7.4). Chlorophyll, certain vitamins and coenzymes, and a host of visual and body pigments (especially carotenoids) fluoresce. Fluorescence is also seen in the cuticle of many arthropods (e.g., euphausiid shrimp and scorpions), though it is not the chitin itself that fluoresces, but a minor molecular component. While many fluorescent biological molecules have excitation and emission spectra in the ultraviolet range (and are thus less likely to serve a visual function), there are still many in which both spectra are primarily in the visible range. However, even conspicuous fluorescence may often be an innocuous and unavoidable side effect of light absorption. Remember that, before all else, a fluorophore is an absorber. So, just because certain scorpion species blaze green or blue under UV light, it doesn't mean that the fluorescence serves an ecological function.

The ubiquitous and often irrelevant nature of fluorescence is important to remember, because it is easy to get carried away once you see a fluorescent image of an animal. As I mentioned earlier, most animal fluorescence is detected by illuminating the specimen with bright short-wavelength light (usually near-UV and blue) and then viewing it through a filter that blocks the illumination but not the emission wavelengths. While a good way to detect and map fluorescence, this method can exaggerate its importance. For example, one of the most fluorescent "natural" substances I have seen is a Froot Loop. Hit a handful of these with blue light and view them through a filter that blocks the blue and I guarantee you will be impressed (and maybe never eat them again). However, under normal illumination, they don't look spe-

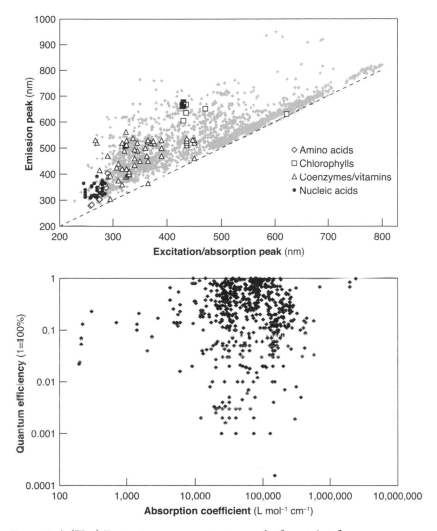

Figure 7.4: (Top) Excitation versus emission peaks for ~3600 fluorescent materials that have clear single peaks in both spectra (a mixture of natural substances and commercial fluorophores). The dotted line denotes where the two peaks are at the same wavelength. Note that all the data are above this line, though some of them are quite close to it. (Bottom) Absorption coefficient versus quantum yield (aka "fluorescence efficiency") for ~700 substances for which both are known (mostly commercial fluorophores). The brightness of a fluorescent substance is proportional to the product of the absorption and the efficiency, so the substances in the upper right corner are the brightest. Data from McNamara et al., 2006.

cial. Even under blue light (without the viewing filter), all you will notice is that they are slightly oranger and redder than they should be under blue light. The same is true of nearly all fluorescent animals. Even strikingly fluorescent corals look only a bit more colorful than you would expect when viewing them at depth without special filters. In fact, you would likely miss the extra color if you didn't know what to look for.

This all suggests a few things. First, as always, don't fall in love with the pretty pictures you make in the lab. There is no substitute for going into the field and seeing an organism under natural lighting. Second, fluorescence is most likely to have an ecological purpose when it occurs in an environment with bright, nearly monochromatic, short-wavelength light. The two main cases of this are clear water between the depths of 10 meters and 100 meters, and blue-emitting photophores. Longer wavelength monochromatic illumination works as well, but because the fluorescence has an even longer wavelength and should be fairly separated from the excitation to be of any real benefit, you run out of visible spectrum sooner. So look for it in coral reefs, clear lakes, the epipelagic zones of the tropics, and in photophores. Third, for fluorescence to be seen with any real contrast, the viewer has to be less sensitive to the shorter wavelengths of light in the environment. As we discussed in chapter 4, visual pigments have broad absorption spectra, so the best way to limit short-wavelength sensitivity is to have a filter that blocks short-wavelength light. Therefore, look for cases in which the viewer of the fluorescence has a filter, which will most likely look yellow. Interestingly, the lenses and corneas of a fair number of coral reef fish are yellow, suggesting that they may be optimized for seeing fluorescence (Marshall and Johnsen, in prep.).

So, which functions have been hypothesized and which have been demonstrated? Probably the only case in which fluorescence has been fairly conclusively shown to play a functional, optical role is in bioluminescence. Many photocytes, particularly those in cnidarians, use green-fluorescent protein to convert the spectrum of their emitted light from blue to green (Haddock et al., 2009) (figure 7.5). Given that coastal waters transmit green light farther than blue light and that the bioluminescence of coastal species (including cnidarians) is primarily green (Haddock et al., 2010), it is possible that this is done to increase visibility. Also, while deep-sea species are more sensitive to blue light, coastal and shallow-water species are more sensitive to green light, making green bioluminescence a more visually effective signal. Also, as we discussed in chapter 3, fluorophores appear to be used in some biolumines-

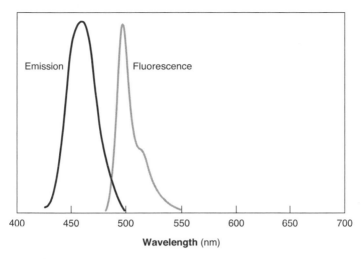

Figure 7.5: Light emission of the bioluminescent hydromedusa *Aequorea victoria* with and without the presence of green fluorescent protein. The original emitted light peaks at about 470 nm, but then is converted to greener light by the co-occurring GFP. Data from Haddock et al., 2010.

cent species to make red light from shorter-wavelength light. One can, of course, quibble about any adaptive argument, but I would be willing to bet a small farm that fluorophores in photophores are there to modulate the spectrum of the emitted light.

Once you get away from photophores, things become less clear. Perhaps the most brightly fluorescent taxa (at least in the marine world) are the corals and anemones. Reef-building corals in particular are quite fluorescent, nearly all of them displaying GFP-related green and red fluorescence (Matz et al., 1999). However, corals are not bioluminescent, leaving the function of the GFP fluorescence obscure. The two main hypotheses are that the GFPs either (1) provide photoprotection by absorbing short-wavelength photons and disposing of some of the energy via less energetic and thus less damaging photons, or (2) enhance photosynthesis for the symbiotic zooxanthellae by converting the incident light to wavelengths that are more efficiently absorbed. Both hypotheses are plausible, but neither has been proven. In fact some evidence against both has been found (reviewed by Mazel and Fuchs, 2003).

Moving away from corals, only a few studies have examined the potential of fluorescence as a signal. For example, several species of stomatopods have fluorescent yellow markings on their antennal scales (Plate 7) (Mazel et al., 2004). We discussed stomatopods in chapter 3 in relation to sonoluminescence and will discuss them again in chapter 8 regarding their ability to see linearly and circularly polarized light. The fluorescent antennal scales of the stomatopod *Lysiosquillina glabriuscula* are presented to both conspecifics and predators in a common threat display. The animal is found in a monochromatic, blue environment, and its twelve-pigment visual system is strongly filtered, so that the contrast generated by the fluorescence is substantial. Calculations by Mazel et al. (2004) showed that the fluorescence brightens the patches by about 30% (as viewed by the animal). This is all good circumstantial evidence that fluorescence plays a role, but so far it has not been confirmed by experiment (at least nothing that is yet published—stay tuned).

Fluorescence has also been hypothesized to function as a signal in fish. A recent study of red fluorescence in coral reef fish (Michiels et al., 2008) found thirty-two species from sixteen genera that displayed strong red fluorescence under filtered lighting. In general, the fluorescence appeared to be due to a substance that co-occured with the guanine crystals in their scales. However, with few exceptions, the fluorescent regions were not conspicuous (or even red) under broad-spectrum illumination. Unfortunately, photos of the fish under ambient underwater illumination without filters were not presented, so it is difficult to determine how much the fluorescence affects the fishes' appearance in the wild. One of the fish, the goby *Eviota pellucida*, has a visual pigment that would make it sensitive to its own red fluorescence. However, since the visual pigments of the other fish were not tested, it is difficult to know whether this is a fluke, since red sensitivity is not especially rare in near-surface fish.

By far the best evidence for fluorescence as a signal enhancer comes from a clever study on the parrot *Melopsittacus undulates* (Arnold et al., 2002). The crown and cheek feathers of this bird, which are used in courtship displays by both sexes, are highly fluorescent, with an excitation spectrum in the near-ultraviolet and a broad emission spectrum peaking in the yellow. Calculations showed that this fluorescence made the plumage about 14% brighter to the visual system of these animals. To determine whether this had any effect on courtship, Kathryn Arnold and her colleagues Ian Owens and Justin Marshall smeared sunblock on the birds' heads (controls were smeared with petroleum jelly, which does not appreciably absorb UV light). This removed the UV excitation illumination and thus the fluorescence. However, because the feathers reflected essentially no UV light, it did not affect their appearance at these wavelengths. They found that, when viewing the opposite sex, both males and females strongly preferred individuals without the sunblock. However, when viewing the same sex, no preference was found. The birds could not smell each other, so it is unlikely that the odor of the sunblock had an effect. Together, this all suggests that the fluorescence of the plumage is important in mate choice.

Given the simple nature of this last experiment, you'd think it would have been tried in other animals over the last eight years, but sadly these few paragraphs review the entirety of what is known about the ecology of animal fluorescence. I have participated in a few research cruises with Mikhail Matz, during which he found a number of suggestive examples of fluorescence, including fluorescent sharks, bright fluorescent eyes in fish, and a copepod in which the male's antennae fluoresce brightly. All tantalizing, but nothing is known for sure yet. The ecology of fluorescence is still very much a scientific frontier.

Phosphorescence and Inelastic Scattering

Two processes hover at the edge of the field of fluorescence, one commonly known and possibly biologically irrelevant, the other barely known but possibly important. The first is phosphorescence. What is and what is not phosphorescence can get muddled, so I would like to first lay it out in terms of things you see in everyday life. Glow-in-the-dark stickers and stars are phosphorescent, black-light posters sometimes are but often are not (they are usually fluorescent). Glow sticks are not phosphorescent, and neither of course is bioluminescence (both are examples of chemiluminescence). Basically, if a substance requires an initial exposure to light in order to glow, but will continue to glow for a while after the light is turned off, it is phosphorescent. This rule doesn't work in the other direction, though. There are phosphorescent substances that glow for only milliseconds after the exciting light is turned off. The red, green, and blue phosphors in a (now) old-style color television set are a good example. The phosphors lining the inside of "fluorescent" bulbs that convert the primarily ultraviolet original emission to visible light are another.

As long as the substance glows for at least one millisecond after the exciting light is turned off, it is referred to as phosphorescence. This might seem awfully fast, but fluorescent substances generally stop glowing less than one nanosecond after the lights are turned off, so phosphorescence is at least a million times slower. I really cannot do justice to the explanation of why some substances are phosphorescent and others are merely fluorescent without writing a separate book on quantum mechanics, so what follows is the nickel explanation. As we discussed at the beginning of this chapter, in fluorescence a molecule is excited by a photon to a higher energy state, drops back down in many small jumps and one final larger jump that emits a photon of longer wavelength. In phosphorescence, the same thing occurs, but the molecule gets stuck at a slightly lower energy state before it can execute the last jump and emit the photon (figure 7.6). It's stuck because, somewhere along the way, an electron flipped its spin. "Spin," like most quantum mechanical terms, does not actually mean spin. The electron is not physically spinning and the property of "spin" can only take two values; up or down. If you took college physics, you may remember something called the "Pauli Exclusion Principle" that says (roughly) that two electrons cannot have the same orbit and spin. So, if the excited electron flips its spin, it can't drop back down to its old orbit, since there is already another electron there with that spin. The transition is said to be forbidden and the excited electron is trapped until the spin flips back and it can drop to the ground state and emit its photon. The spin flipping takes time, anywhere from milliseconds to minutes, so the substance continues to glow after the exciting light is turned off. I have just run rough over decades of subtle physics, but I hope this explanation gets the point across.

While phosphorescence looks like it could be of value to animals, especially those active at dusk or moving between bright and dark environments, to my knowledge there are very few biological examples of this process. It is possible that somebody has documented biological phosphorescence in the more distant past, but searching for it is next to impossible because "phosphorescence" was used interchangeably with "bioluminescence" until recently. In fact, the only solid evidence for true phosphorescence that I have seen is found in scleractinian corals. It has been known since the mid-1980s that cross sections of coral skeletons viewed under UV light display alternating yellow/green and blue bands that correspond with their annual growth cycles. Later, Wild et al. (2000) showed that about 10% of the fluorescence

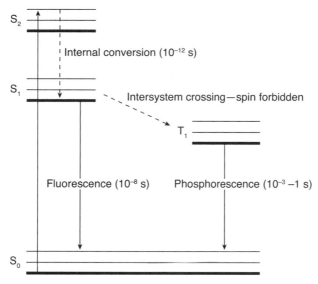

Figure 7.6: Energy diagrams of fluorescence versus phosphorescence. S0 is the ground state and S2 is the initial excited state. In normal fluorescence, the energy drops to S1 without emitting a photon and then drops from S1 to S0 while emitting a photon. In phosphorescence, the S1 state is converted to a different state T1 that has a flipped spin and thus cannot quickly execute the jump to the ground state and emit the photon. Note also that the forbidden state is slightly lower, meaning that phosphorescence is at slightly longer wavelengths than fluorescence.

of the yellow/green bands was actually due to phosphorescence that had a decay time of 1.5 seconds. It is not known what the phosphorescent substance is, and the emitted light is highly unlikely to play an ecological role.

It is difficult to know why phosphorescence appears to be rare in organisms. One possibility is that people simply have not looked for it. The problem with phosphorescence is that you have to illuminate the substance with a bright light and then quickly turn out the light and look for a dim signal. Human visual systems do not dark-adapt quickly, so it could be easily missed. This fact also suggests an ecological reason for its apparent rarity—animals seldom go from bright regions to dark ones in a hurry, and, if they do, their eyes are likely not dark-adapted. Finally, it may actually be a matter of constraint. Unlike fluorescence, which occurs in a host of biological molecules, phosphorescent molecules tend to include biologically unusual elements like zinc, strontium, aluminum, selenium, cadmium, and various rare earths.

While organisms do contain these substances, it is generally in very low concentrations. Many are also toxic.

My hunch is that, if phosphorescence occurs anywhere, it occurs in photophores and is used to lengthen the duration of light emission. Even there, it is likely to be rare, because much bioluminescence is rapid and long-duration bioluminescence is usually of bacterial origin (and thus constant). However, it would be interesting to look for!

The other fluorescence-like process is inelastic scattering. As I mentioned in chapter 5, light scattering does not really involve a photon bouncing off a molecule. Instead, it might be better to say that it involves absorption followed rapidly by emission. In this way, fluorescence, where a photon is absorbed and one of less energy is emitted in a random direction, could be considered a case of inelastic scattering (a collision in which energy is lost). However, the term "inelastic scattering" is reserved for other processes. Most of these involve aspects of physics that are unlikely to occur within or near organisms, but one—known as "Raman scattering"—may turn out to be ecologically relevant.

As with fluorescence, in Raman scattering a photon strikes a molecule, and then a photon of different energy leaves the scene. Also like fluorescence, the energy of the new photon can be higher than before, but is usually lower, with the exact emission spectrum being diagnostic of the encountered molecule. This is where the similarities end though. The big difference is that there is no fixed emission spectrum. As we discussed earlier in the chapter, a fluorescent substance has a set emission spectrum that is independent of the wavelength of the exciting light. In Raman scattering, the excitation wavelength determines where the emission spectrum will fall. If you shift the excitation light from a 400 nm laser to a 500 nm laser, the emission spectrum will shift by about 100 nm (the shift is exact if you work with frequencies). The other difference between fluorescence and Raman scattering is that the latter is far weaker. Generally less than one in ten thousand of the photons gets converted to a photon of a different energy.

Because it is such a weak phenomenon, it is only likely to have an ecological effect if several conditions are met. First, there must be a source of bright, nearly monochromatic light to create a detectable number of Raman-scattered photons and to easily separate them from the excitation light. Second, there must be a huge volume of transparent material to give the photons the best chance of becoming Raman scattered. Both of these

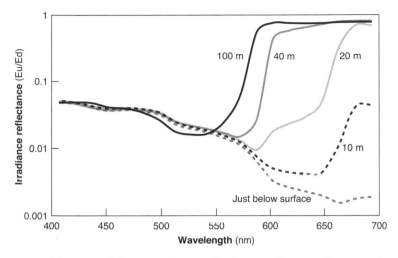

Figure 7.7: The ratio of the up-welling to the down-welling irradiance in clear ocean water (1 means that the two irradiances are equal). The dramatic increase in this ratio at longer wavelengths at depth is due to the presence of Raman-scattered light.

conditions exist in the upper 100 meters of the open ocean, and it is here that Raman scattering is most likely to have a significant ecological effect. In fact, Raman first begin thinking about this process during a sea voyage between London and Bombay where there was little else to do but ponder the deep-blue color of the sea.

Though Raman scattering in water has been known for quite some time, it was familiar to only a subset of physicists and optical oceanographers. I stumbled across it while studying cryptic coloration in pelagic animals in 2000. My plan was to predict what the perfect cryptic colors would be for animals as a function of their depth and other factors. In order to do this, I needed to know how much light there was in every direction at a number of wavelengths. I first did this using a sophisticated optical modeling package. It went well until I noticed two things: (1) there was much more long-wave-length light at depth than I expected there to be, and (2) this long-wave-length light seemed to be evenly distributed in all directions. In other words, there was just as much long-wavelength light going up as going down (figure 7.7). This bothered me because any scuba diver knows that down-welling light is far brighter than up-welling light. I assumed I had done something wrong and tracked down some real measurements of down-welling and up-

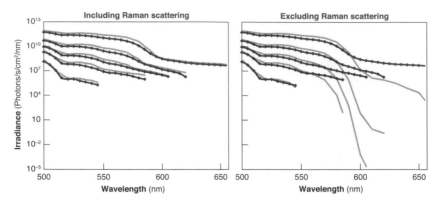

Figure 7.8: Measured (lines with points) and modeled (thin lines) down-welling irradiance at 50 m, 100 m, 150 m, 200 m, and 300 m depth in the open ocean. The modeled spectra that exclude Raman scattering show far less long-wavelength light than the measurements.

welling light in the ocean. These measurements had the same two problems. There was more long-wavelength light than I expected and nearly as much going up as going down. So I started talking to various experts in ocean optics and learned about Raman scattering.

What I learned is this. First, as you go deeper into the ocean, long-wavelength (orange and red) light gets absorbed quickly, and very little is left after a few meters, even in the clearest of waters. You would expect that if you went even deeper the long-wavelength light would continue to be absorbed and would be undetectable by any means. However, as these "solar" long-wavelength photons are being absorbed, new ones are being created via Raman scattering. Raman scattering is a weak process, but the blue light is so much brighter than the long-wavelength light, and there is so much water, that many more long-wavelength photons are created than were already present. In fact, instead of attenuating very quickly like it should in water, the long-wavelength light attenuates at about the same speed as the blue light, because the blue light is what is making the long-wavelength light. So you end up with far more long-wavelength light at depth than you would expect. I once did the calculations and, by the time you get to 1000 meters, the ratio of Raman-created red light to the few original red photons from the actual sun is about two hundred orders of magnitude (figure 7.8). The isotropic nature of this long-wavelength light (its equal brightness in all directions) comes from the fact that the Raman effect scatters light equally in all directions.

It is critical to realize that the deep-sea light field is not red, or even orange. It is still overwhelmingly blue, at least to our eyes, and the blue light is several orders of magnitude brighter than the orange and red, even with Raman scattering. However, remember that many animals, including us, can detect light that varies in intensity over about ten orders of magnitude. So, the Raman-scattered light is bright enough to be seen if the visual system can filter out the much brighter blue light. In some ways, this is no different than detecting weak fluorescence. In other words, you need an excellent filter that blocks the blue light, while still letting through the longer-wavelength Raman scattered light. This sounds challenging, but the human lens has an optical density of about 4 to 360 nm UV light, compared to nearly zero for 550 nm yellow light (Griswold and Stark, 1992) This means that the lens is blocking 99.99% of the 360 nm light, which is about what you would need to filter out enough of the blue light to see Raman-scattered light in water. So it is possible.

However, why bother? The main reason is because the light field of Raman-scattered light has such a different shape from the usual light field. Because the illumination is nearly the same in every direction compared to the usual dominance of overhead light, the optimal strategies for camouflage are different (Johnson, 2002). I won't go into all the geometric arguments, but the bottom line is that hiding in uniform light fields requires very high reflectances, close to 100% in fact. Most animals have much lower reflectances, which are optimal for hiding in the usual "brightest from above" light field. Therefore, an animal with the correct filter may be able to spot animals that are otherwise hidden.

CONCLUSION

Organismal fluorescence is a fascinating and charismatic topic about which not enough is known. Once you step away from the voluminous databases of commercially available fluorophores and phylogenetic analyses of GFPs, you are in a land with few answers and many questions, some of which could be answered via simple experiments. Honestly, I don't know why more people aren't working in the field.

FURTHER READING

There are endless books on fluorescence microscopy, but none on the natural functions of fluorescence or its prevalence in organisms. There are also no review articles. The same is true for phosphorescence and Raman scattering. However, books on pigments and bioluminescence often discuss fluorescence in passing.

CHAPTER EIGHT

Polarization

> Don't undertake a project unless it is manifestly important and
> nearly impossible.
> —EDWIN LAND (from "The Vindication of Edwin Land," *Forbes*)

Edwin Land was a remarkable person. Though he never finished college,
Land was a dominating figure in photography and optics for much of the
twentieth century and made fundamental contributions to the understanding of color vision. His two most famous inventions were the Polaroid camera and the polarizing filter. The latter, developed late at night using equipment and space stolen from Columbia University, opened the door for a
myriad of technical and scientific applications, ranging from 3D movies to
the study of honeybee navigation. So, it is with his inspiring, and somewhat
intimidating, example that I tackle the issue of polarized light.

As with radiometry, polarization can be a confusing topic. Unfortunately,
unlike radiometry, its complexity is not primarily due to confusing units.
The physics of polarized light is genuinely tricky. This is another subfield of
optics that is made easier by thinking of light as a wave. You can think of individual photons as having polarization and work from there, but—believe
me—you don't want to.

Because polarization has often been described poorly, I have started this
chapter with what I hope is a clean introduction to the subject. It is a bit abstract, but bear with me. We do get to the interesting phenomenology of the
subject (dancing bees and rolling dung among others) later in the chapter.

KINDS OF POLARIZATION

Remember that light can be thought of as a wave that propagates through
space. Now imagine a perfectly monochromatic, perfectly collimated beam

of light in a perfect vacuum (optics is full of platonic ideals). If you hold a small electric field meter in the path of this beam, it shows that the magnitude of the local electric field changes very rapidly in a sinusoidal fashion (we are assuming that the meter can keep up with the rapid changes, which it likely cannot). But what about the direction of the field? Unlike air pressure in a sound wave, electric fields are vectors and thus have both magnitude and direction. This direction is, for our purposes, always perpendicular to the direction of the propagation of the light beam and is called the "e-vector." The e-vector is denoted by an angle, usually relative to the vertical (though any reference will do as long as you are consistent) (figure 8.1).

You may notice that I left out the magnetic field. Because the magnetic field is usually (but not always) perpendicular to and in phase with the electric field, knowing the e-vector tells you the magnetic field vector. In most cases, the electric field has a far stronger effect on matter, so the convention has been to describe polarization by e-vector, though scientists in some countries prefer to use the magnetic field vector.

Like all vectors, the e-vector can be broken down into two perpendicular components that can be at any angle, though it is often easiest to just use horizontal and vertical, which we'll do here. Thus, you can consider any monochromatic parallel beam of light of as the sum of two beams, one with a horizontal e-vector and one with a vertical e-vector. Of course, just because you can do something doesn't mean you have to. What do we get in return? A tidy way of looking at polarized light, it turns out.

Like all simple waves, the horizontal and vertical components of the e-vector are completely described by three things: frequency, amplitude, and phase. We said that our beam is monochromatic, so we can ignore frequency. In fact, let's just set it to one. We don't care about the absolute phases either, just the phase difference between the two beams. So we'll set the phase of the horizontal component to zero. After these simplifications, the horizontal and vertical components of the electric field measured by our little meter are:

$$E_h = A_h \sin(t), \text{ and } E_v = A_v \sin(t + \phi), \qquad 8.1$$

where t is time, A_h and A_v are the amplitudes of the horizontally and vertically polarized components, and ϕ is the phase difference between the two. What happens to the electric field over time? If the phase difference ϕ is zero, then $E_v = \dfrac{A_v}{A_h} E_h$, which means the amplitude of the vertical field is just a

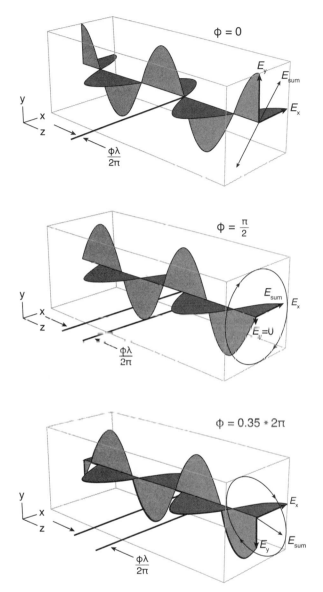

Figure 8.1: Linearly, circularly, and elliptically polarized light broken into vertical and horizontal components.

multiple of the horizontal field, so the angle of the e-vector never changes, except to go in the opposite direction for half the cycle. If A_v is zero, then E_v is also always zero, so the e-vector is horizontal. If A_h is zero, then the e-vector is vertical. If they are both zero, then there is no light at all. In general, the angle of the e-vector is the arctangent of $\dfrac{A_h}{A_v}$.

We have just described linearly polarized light. Before we move on, there is a small point to deal with. E-vector angles only go from 0° to 180° rather than from 0° to 360°. This is because they point in opposing directions, depending on the current phase of the wave. Thus, it is understood that the e-vector angles θ and $180° + \theta$ are the same. In fact, in common usage, the e-vector is not a vector, but an axis. Names are never quite what you would like them to be. Some people prefer to use "angle of polarization" or the more accurate term "axis of polarization," but "e-vector" dominates scientific papers so we'll stick with it. However, we'll add "axis" to make it clear that it points both ways.

I know it looks like we just used a cannon to open a door that was already open, but this is because linear polarization is only a limiting case of something more general. Suppose the phase difference ϕ in equation 8.1 is not zero. Then, the endpoint of the electric field vector, instead of just telescoping in and out and backward on one axis, traces out a curve. In other words, the axis of the e-vector changes over time, but in a predictable way. In general, the curve traced is an ellipse whose shape and orientation depends on the phase difference between the two components and their relative amplitudes (figure 8.1). In geometry, a line can be considered a flat ellipse, so the axis you get in linear polarization is still an ellipse. Think of it as a limiting case.

The other limiting case of an ellipse is a circle. This happens when the $A_h = A_v$ and the phase difference between the two components is either 90° or -90°. This special case, when the electric field vector traces out a perfect circle 10^{14} times a second (for visible light), is called "circular polarization." If the phase difference is 90°, then the polarization is called "right-handed," because, if you could stop time, the helical path of the e-vector would look like the threads on a typical right-handed screw (it is sometime also called "clockwise"). If the phase difference is -90°, the polarization is called "left circular" or "counter-clockwise." Unless you are an electrical engineer, and then the naming convention is opposite. Frankly, every time the terms "right-handed" and "left-handed" show up in physics and engineering, my head starts to ache. Usually, you look up which is which for your particular system, forget it, look it up again, forget it, and so on.

Circular and elliptical polarization may seem esoteric. Who cares if the electric field vector traces a circle or ellipse 10^{14} times a second? Who can even see it? It turns out, though, that circularly polarized light is found in nature, and that some animals can indeed see and make use of it. More on this later. First, we need to exchange the idyllic and fictional world of perfect waves for choppy reality.

PARTIAL POLARIZATION

Ideally, the polarization of light is simply defined by its e-vector, which either lies on a single axis or traces out an ellipse. But monochromatic parallel beams do not exist in nature (or in the lab either, if you want to be picky). So what is the polarization of a real beam of light—for example, from the sun— that is the superposition of many waves of different amplitudes, wavelengths, and e-vector angles? If our electric field meter could measure the e-vector of sunlight in real time, it would find that it was varying wildly. In fact, the e-vector of unpolarized visible light changes so quickly that no known instrument can measure it.

In many cases, direct sunlight being a good example, the e-vector averaged over any biologically relevant time period is random. Light of this type is called "unpolarized" and is easy to deal with —we just forget that polarization even exists. However, the e-vectors of many important sources of light—skylight, reflected light, underwater light—are not entirely random. Light from these sources is said to be partially polarized, because, while the e-vector isn't on one axis or cyclic, it is statistically biased toward certain angles and/or handedness. There are, of course, an infinite number of statistical distributions of e-vectors, so you might imagine that characterizing partially polarized light would be impossible. Happily, though, you can completely characterize the polarization of any beam of light (at a given wavelength) using a commercial light meter and two inexpensive pieces of plastic.

The first piece of plastic is a typical polarizing filter. These filters, found in Polaroid sunglasses, LCD monitors, and many other places, are so ubiquitous that we forget what a marvel they are. There have been several versions over the years, but they are mostly based on parallel arrays of polymers whose absorption of light depends on the axis of the e-vector relative to the long axis of the polymer. Substances like this are known as "dichroic" (a word that unfortunately has at least two other optical meanings). When the e-vector and polymer axes are parallel, the light is absorbed strongly. When they are perpendicular, the light is hardly absorbed at all (we call this the "transmission axis of the filter"). For all other angles, the fraction of light absorbed depends on the square of the cosine of the angle between the polymer and the e-vector axes. So a perfect polarizing filter will absorb none of the light that has an e-vector perpendicular to the polymer axis and all of the light that has an e-vector parallel to it. No polarizing filters are perfect, of course, but modern ones come quite close, though only over a limited spectral range. Their quality is described by what is called the "extinction ratio," which is the ratio of the amount of linearly polarized light that passes though the filter when the polymer and e-vector axes are perpendicular to the amount of light that passes though the filter when the axes are parallel. Typical polarizing filters have extinction ratios of at least 1000:1—impressive for a piece of plastic that costs only a few dollars.

The second piece of plastic we'll need is a circular polarizing filter. Just like a linear polarizing filter, it only lets through light of one polarization, but in this case it separates right-handed from left-handed circularly polarized light. Despite what you might think, it is not made of helically wound polymers, though it could be. In fact, DNA works as a circular polarizer, as do the helically arranged polysaccharides in the cuticle of certain arthropods and certain secondary structures in folded proteins (e.g., Neville and Caveney, 1969). Commercial circular polarizers however, are made using a different trick that we'll discuss later. For now, just assume you can get one for a few tens of dollars.

So how do we use a linear and circular polarizer to characterize the partial polarization of a beam of light? First, we place the light meter in the path of the beam (facing the source) and measure the irradiance (at a given wavelength), which we'll call I. Nothing mysterious here, we are just measuring how bright the beam is. Then, leaving the light meter where it is, we put the linear polarizer in the beam with its transmission axis oriented horizontally and take a measurement. Then turn the polarizer 90° (so that it now maximally transmits vertically polarized light) and take another measurement. Subtract the third measurement from the second and call this number Q. If the light is unpolarized, it doesn't matter how the polarizer is oriented and Q equals zero. If the light is completely horizontally polarized, Q will equal about $I - 0 = I$ (it won't exactly equal I because no real polarizer is perfect). So it is tempting to conclude that Q/I tells us how polarized the light is. But what if the e-vectors of the beam were all oriented parallel to the $45°-215°$ axis? (or the $-45°-135°$ axis)? Then each polarizer would let through half the light ($= \cos^2[45°]$) and Q—being the difference of the two—would equal zero even though the light was completely linearly polarized. How do we get around this? We take two more measurements, one with the polarizer at $45°$ and one with the polarizer at $-45°$ and again calculate the difference, which is called "U."

If we ignore circular polarization (which we often can do in biology), then I, Q, and U completely describe the polarization of any beam of light, whether it is totally or partially polarized. In other words, measuring what fraction of light is transmitted through a simple linear polarizer with its transmission axes oriented at $0°$, $90°$, and $\pm45°$ tells us everything we need to know about the linear polarization of even the messiest signal. Even more amazing, to my mind, is that we needn't know all four measurements, just their sums and differences. The multidimensional monster under the bed turns out to be just three numbers.

The irradiance I and the two differences, Q and U, are three of the four Stokes parameters, described by Sir George Stokes in 1852, decades before the electromagnetic theory of light was even developed. They are a wonderful example of the reduction of a complex phenomenon to a few simple rules. Unfortunately, in their raw form, Stokes parameters are not typically useful to most biologists. In most cases, biologists who study polarized light are interested in two things: the degree of polarization and the average angle of the e-vector.

This is where things get even easier, because it turns out that any beam of partially polarized light can be considered the sum of two beams, one completely polarized and one completely unpolarized. The e-vectors in the unpolarized beam are random and those in the polarized beam trace out the simple ellipse we discussed above. I find this amazing. I have a hard time dividing my somewhat messy desk into a perfectly orderly desk and a completely chaotic one, even in my mind.

Even though there isn't a device that will actually let you split the light beam in this way, the fact that you can mathematically do it makes things easier. For example, the degree of linear polarization p_{linear} is then just the fraction of the light beam that is totally polarized. There are two ways to calculate this. First, using the Stoke parameters:

$$p_{linear} = \frac{\sqrt{Q^2 + U^2}}{I} . \text{ (We are still ignoring circularly polarized light.)} \quad 8.2$$

You can see from this that the horizontal-vertical and 45°polarizations add up in a "Pythagorean Theorem" sort of way. However, there is an even simpler way to measure the degree of linear polarization, if you are willing to watch the light meter as you rotate the polarizer. Just rotate it until the irradiance at the light meter is maximal—call that value I_{max}. Then rotate it 90° (it doesn't matter in which direction). The irradiance at the light meter will now be as low as it can get—call it I_{min}. The degree of polarization is then:

$$p_{linear} = \frac{I_{max} - I_{min}}{I_{max} + I_{min}} . \quad 8.3$$

This method is quick and also gives you the e-vector axis, which is the just the angle of the transmission axis of the polarizer at I_{max}. However, in many cases you are not free to rotate the polarizer. For example, people who do polarization imaging with automated cameras set them up to rotate through the four canonical positions and use equation 8.2 to calculate the degree of polarization. The angle of the e-vector axis of the polarized component is then given by:

$$\tan 2\theta = \frac{U}{Q} . \quad 8.4$$

Equation 8.4 looks harmless, and you might be tempted to just solve for θ. But, if you remember your high school trigonometry, you will also remember that arctangents are a real pain because you have to worry about which quadrant of the circle you are in. There are actually three solutions, depending on the signs of U and Q, which are given in appendix F.

Instead of working with Stokes parameters, you can also just use the irradiances measured with the polarizer at three angles to get the axis of polarization. In this case:

$$\tan 2\theta = \frac{I_{90} + I_0 - 2I_{45}}{I_{90} - I_0}.$$ 8.5

Again, you have to worry about quadrants (see appendix F for the three solutions). It's straightforward, but tedious enough that you can see why people prefer to rotate a polarizer until the maximum amount of light is transmitted.

What about circularly polarized light? This is where the circular polarizer comes in. Put it in the light beam so that it maximally transmits right-handed circularly polarized light. Call the irradiance at the detector I_{rh}. Then flip the polarizer over so it maximally transmits left-handed circularly polarized light and call the irradiance at the detector I_{lh} (this flipping-over trick only works with certain kinds of circular polarizers; we assume we have one for simplicity). The final Stokes parameter V is just the difference of these two measurements.

I have no idea why the parameters are called "I," "Q," "U," and "V," aside from the obvious I for intensity or irradiance. Stokes actually used $ABCD$ and others have used $IMCS$. My hunch is that IQUV is based on Latin roots, Q being related to quadrat, and V being related to "volute," the Latin word for spiral.

Circularly polarized light doesn't have an angle, but it does have a handedness and a degree of polarization. The handedness is the sign of V (positive means right-handed) and the degree of circular polarization p_{circ} is:

$$p_{circ} = \frac{V}{I} = \frac{I_{rh} - I_{lh}}{I_{rh} + I_{lh}}.$$ 8.6

Another convenience is that you can consider any partially polarized beam of light to be the "Pythagorean" sum of an unpolarized beam, a perfectly linearly polarized beam, and a perfectly circularly polarized beam. In other words, the total degree of polarization p is:

$$p = \sqrt{p_{linear}^2 + p_{circ}^2}.$$ 8.7

This is all you need to know, just a few simple equations, most of which are redundant. With two pieces of plastic, a few quotients, and the occasional square root, you can completely describe the polarization of any light. The next sections deal with why you should bother.

SOURCES OF POLARIZED LIGHT

Direct sunlight, light in murky water, airglow, fire, lightning, fluorescence, triboluminescence, soniluminescence, and direct light from the full moon and the outer planets are unpolarized. The solar corona; skylight; rainbows; halos; zodiacal light; comets; underwater light in the ocean; reflected light; the red aurora; emission from hot metals, lava and tourmaline; and the light from the non–full moon and the inner planets are sometimes polarized. With two exceptions (which we'll get to in a moment), what distinguishes the second group from the first is the presence of scattering.

Polarized light in nature is a scattering phenomenon. However, not all scattering is equally effective at polarizing light. Two kinds work best. The first is single scattering by particles much smaller than a wavelength of light. Remember equation 8 from chapter 5 that showed how small particles scattered light?

$$S(\theta) \propto \frac{V^2}{\lambda^4}(m-1)^2\left(1+\cos^2\theta\right) \qquad 8.8$$

I added the $1 + \cos^2\theta$ term without any explanation, saying I would get to it later. Well, the future is today, as they say.

This equation is actually the sum of two terms, one for each of two perpendicular polarizations of the incident light. This is because the polarization of the incident light affects how a small particle scatters it. Imagine a beam of light passing from left to right and then hitting a small particle glued to the page (figure 8.2). For one polarization, called "parallel" because it is parallel to the page, the light is scattered mostly forward and backward and not at all directly to the side. The equation is:

$$S_{\parallel}(\theta) \propto \frac{V^2}{\lambda^4}(m-1)^2\cos^2\theta. \qquad 8.9$$

For the other polarization, called "perpendicular" because it is perpendicular to the first polarization, the light is scattered equally in all directions:

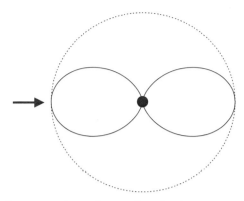

Figure 8.2: The polarization dependence of Rayleigh scattering. The incident light enters from the left. The polarization component with an e-vector that is parallel to the page is scattered primarily forward and backward (solid line). The other polarization component is scattered equally in all directions (dotted line).

$$S(\theta)_\perp \propto \frac{V^2}{\lambda^4}(m-1)^2.$$ 8.10

Don't worry too much about the equations. Also, don't worry about "parallel" and "perpendicular"—they are part of a naming convention that seldom shows up in biology. The important thing to realize is that, when the angle θ equals 90°, all the scattered light will have the same e-vector axis. In other words, it will be completely polarized. Light scattered at other angles will be partially polarized, except for light scattered directly forward or backward, which will be unpolarized because the cosine of 0° and 180° equals one.

As we discussed in chapter 3, the blue sky is the result of single scattering by small particles (in this case, molecules of nitrogen and oxygen), which explains why it is also strongly polarized. As expected, the polarization of skylight is weakest both near the sun and about 180° from the sun and strongest about 90° from the sun (Plate 8). It's not exactly 0°, 180°, and 90°, because our sky is not ideal. The best way to see polarization of skylight is to look up at the sky through Polaroid sunglasses during dawn or dusk. As you rotate your head, a large dark band splitting the sky into a solar and antisolar half will appear and disappear. About 5% of people can see a version of this effect without Polaroid glasses. Instead of dark band, they see a faint yellow hourglass known as "Haidinger's brush." The exact cause of this is unknown, but is thought to be due to dichroism in lutein, a yellow backing pigment in the fovea of the eye (Bone and Landrum, 1983).

The polarization of skylight can be high, but never quite reaches 100%. The finite size of the sun, asymmetry of the molecules in the air, larger particles in the air, and multiple scattering all reduce the maximum polarization to about 80%. Multiple scattering is a major culprit. It is often stated that multiple scattering dilutes the degree of polarization, the implication being that each scattering event is a corrupting process leading a once idealistic and highly polarized photon down a decadent path to unpolarized ruin. However, each scattering event with a small particle creates highly polarized light, but only at 90° to the direction that the photon was first traveling. What we lose in a hazy atmosphere is directionality. When we look at a patch of clear blue sky that is 90 away from the sun, most of the light we see came directly from the sun and then was scattered 90° toward our eye. However, if the sky is hazy, then the patch we look at includes light that came from other places, and thus did not scatter by 90°. It is the addition of this light, along with the reduction of direct sunlight reaching our sky patch, that lowers the polarization.

The e-vector axis of a patch of polarized skylight is roughly perpendicular to a line between the sun and the patch. Therefore, if you can see a few patches of clear sky, you can theoretically triangulate for the position of the sun, even if it is hidden behind clouds or the landscape. This fact opened a whole branch of animal navigation research that we'll get to in a moment.

The same single-scattering process is responsible for underwater polarization, the main difference being that light is primarily scattered by water molecules rather than oxygen and nitrogen. Water is also far denser and murkier than air, with multiple scattering and scattering by water droplets and dust particles (where the polarization of scattered light is lower), so the degree of polarization doesn't usually make it much above 30%, even in clear ocean water. As with skylight polarization, though, underwater polarization is greatest when you are looking 90° away from the main direction of light propagation and minimal when looking either toward or away from it. So, near the surface, polarization is minimal when looking toward or directly away from the sun's image (as refracted by the surface of the water). Farther down, the brightest direction is straight up regardless of the sun's position, so polarization is minimal when looking straight up and straight down and greatest when looking horizontally. Thus, unless you are close to the surface

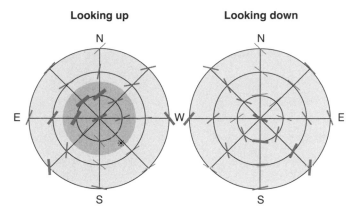

Figure 8.3: Polarization of the underwater light field. The sun can be seen in the southwest quadrant of the "looking up" panel. The thickness of the lines denotes the degree of polarization. From Cronin et al., 2003.

near sunset, the polarization is usually greatest when you are looking horizontally. The angle of the e-vector in this direction is roughly horizontal, with the actual angle depending on the position of the sun and the water's optical properties (figure 8.3).

The other way in which scattering can create polarized light is via coherent scattering, in particular reflection from smooth, substances such as glass, water, and many leaves or structurally colored objects like iridescent butterfly wings. The laws of reflection from surfaces of this kind are a bit complicated (and given in appendix F), but the polarization depends on the angle of the incident light in a fairly simple way (figure 8.4). If incident, unpolarized light hits the surface perpendicularly, the reflected light is still unpolarized. However, as the angle between the light and the perpendicular to the surface increases, the polarization of the reflected light goes up, until it reaches 100% at what is called "Brewster's angle." At incident angles above Brewster's angle, the polarization of the reflected light drops again until it reaches zero when the incident light just grazes the surface. Brewster's angle is easy to calculate; its tangent is equal to m, the ratio of the index of the surface material to that of the surrounding medium. So, in the case of light bouncing off the sea surface, Brewster's angle is the arctangent of 1.33/1, which is 53°. However, don't focus too heavily on the exact angle. The polarization of the reflected light is high over a wide range on each side of Brewster's angle.

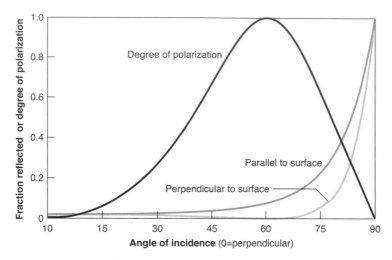

Figure 8.4: Reflection of light from a water surface. The reflection of the polarization component parallel to the surface increases with angle of incidence, but the reflection of the other component drops until it hits zero at Brewster's angle. The degree of polarization of the reflected light is highest at this angle.

So where did the other polarization of the incident light go? The nonreflected beam is refracted through or absorbed by the material, depending on its transparency. This refracted beam is just as polarized as the reflected beam, but with the other polarization. The horizontal polarization is the one that is reflected, which I remember by thinking of a Frisbee bouncing off a sidewalk. Interestingly, both the reflected and refracted beams are completely polarized (at Brewster's angle) when they are traveling in directions that are 90° apart. This 90° rule should remind you of the 90° rule we just discussed in the preceding paragraph on scattering and ultimately stems from the fact that specular (mirrorlike) reflection is actually scattering.

Despite what textbooks often say, light reflected from metals can also be polarized, especially ones that don't conduct electricity well, such as iron and chromium. The degree of polarization only gets up to about 50% and you need to hit the surface at fairly raking angles ($\sim 10°$ from the surface), so it's less often observed, but it's there if you look for it. The only natural exceptions I know to the "scattering is the cause of polarization" rule are the red aurora (polarized by the earth's magnetic field) and light transmitted through dichroic minerals like tourmaline (one of the few dichroic substances of any real size in nature). So unless you are interested in polar illumination on moonless nights or a bug's life in a tourmaline mine, scattering-based polarization covers it. A big open question is whether any animal has linearly polarized bioluminescence. This has only been checked in a handful of species and may be worth further research. Interestingly though, firefly larvae have circularly polarized bioluminescence (Wynberg et al., 1980). It is likely just a side effect of transmission through the helical molecules of the cuticle, but perhaps worth looking into.

BIREFRINGENCE

Before we get to polarized light in biology, there is one last term to deal with. "Birefringence" is one of those nice words derived from Latin that says what it is; it means to have two refractive indices. All substances have different refractive indices for different wavelengths of light. This is called "dispersion" and is the cause of rainbows and the "fire" in cut diamonds. However, a smaller set of substances have significantly different refractive indices for different polarizations of light. Calcite is the classic birefringent substance and fairly common in the biological world, so we'll use it as an example (figure 8.5). A crystal of calcite has what is known as an "optical axis," which is determined by its crystal structure. Light with a polarization parallel to this optic axis experiences a refractive index of 1.486 as it passes through the crystal. Light with a polarization perpendicular to this axis experiences a refractive index of 1.658. This is a big difference, large enough to refract the two beams noticeably differently and create a double image on the other side of a large crystal. In fact, it is so large that, with some clever geometry, you can make one of the two beams head off in a completely different direction and only transmit one polarization. The first polarizers were made this way and the high-quality polarizers underlying DIC microscopy are still large chunks of calcite. Nicol and Wollaston prisms and their various brethren are works of art. However, they are appallingly expensive, so it's a good thing Land invented Polaroid filters.

Aside from building expensive polarizers, what can be done with birefringent materials? Despite what you sometimes hear, birefringent materials themselves do not polarize light, they only transmit the two polarization components at different phase velocities. This difference, though, changes the phase relationship between the two components and thus can convert

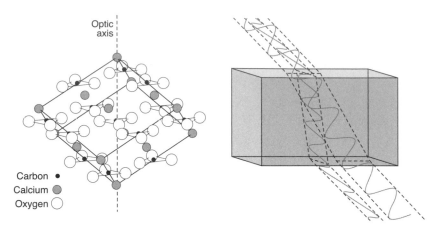

Figure 8.5: (Left) The crystal structure of calcite, showing the optical axis. (Right) Calcite's birefringence. The two polarization components experience different refractive indices within the material and thus are "bent" through different angles as they enter and leave the material. The component with the e-vector parallel to the optical axis is refracted less.

one type of polarization into another. As we discussed earlier, in linearly polarized light, the two perpendicular polarization components have the same phase. If you send this light through a birefringent substance, however, one of the polarizations will get through faster than the other and the phase relationship will no longer be zero. In other words, you will have changed linear polarized light to elliptically polarized light. If the birefringent substance is just the right thickness and you orient it so that the fast and slow axes (low and high index) are both 45° to the e-vector of the incident linearly polarized light, the light that comes out the other side will have phases 90° apart and thus be circularly polarized (see figure 8.6 for an example of what a half-wave plate can do). In other words, you can use this object, known as a "quarter-wave plate" (because 90° is one-quarter of the full 360° cycle of a wave), to turn linearly polarized light (at a given wavelength) into circularly polarized light or vice versa.

Many minerals are birefringent and geologists have used polarization microscopy for decades as a research tool (and to make pretty pictures). Basically, you put a transparent slice of rock between two polarizers that are at 90° to each other. These two polarizers, said to be crossed, should ideally let almost no light through. However, any birefringent substance between them

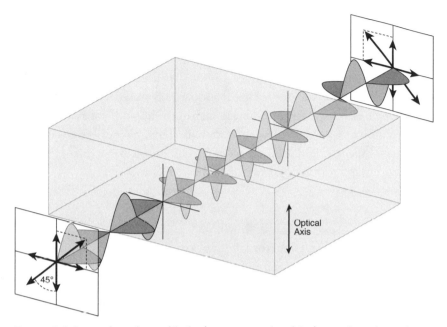

Figure 8.6: Linearly polarized light (e-vector angle of 45°) traveling through a birefringent material. The vertical polarization component travels slower than the horizontal one, and, in this case, the substance is just thick enough to change their relative phases by half a wavelength, which switches the polarization to −45°. A plate half as thick would have converted the linearly polarized light to circularly polarized light.

will convert some of the linearly polarized light to elliptically polarized light or even to linearly polarized light with a different e-vector axis, part of which can then get through the second polarizer. So you see bright areas of birefringent substances on an otherwise black background.

The sort of birefringence that geologists study is usually intrinsic birefringence, which means that it is a property of the crystal structure itself. Intrinsic birefringence exists in animal tissue as well, but more often the birefringence is what is called "form birefringence." Form birefringence is due to the organization of the molecules, in particular long parallel bands of fibers or molecules. Substances like muscle and connective tissue therefore can have substantial form birefringence. Stretching certain plastics can also cause the molecules to line up, creating a form birefringence that is called "stress birefringence." Wear Polaroid sunglasses while driving and you will notice dark spots on your windshield or those of others, each about the size of a small handprint. They are caused by stress/form birefringence due to the processes involved in shaping a flat sheet of glass into a windshield. The source of the polarized light that allows you to see it is usually either polarized blue sky (if it's your windshield you are looking at) or partially polarized reflected light from some nearby glass or shiny object, such as the hood of your car. You can make your own stress birefringence by stretching cellophane between crossed polarizers. I spent a significant chunk of my childhood doing this.

Detection of Polarized Light by Animals

As I mentioned earlier, aside from the few people who see Haidinger's Brush in the sky, the polarization characteristics of light are invisible to humans. However a host of animals can detect one or both aspects of linearly polarized light (see Talbot Waterman's massive review [1981] and Horváth and Varjú's even more massive book [2004] for comprehensive lists of taxa). Arthropods are the big winners here, especially insects, though also most crustaceans and certain spiders and scorpions. In fact, it is unusual to find an insect without polarization sensitivity. Outside of the arthropods, the other major polarization-sensitive invertebrates are the cephalopods. Among vertebrates, polarization sensitivity is rarer and more controversial, but has been found in certain fish (primarily trout and salmon), some migratory birds, and a handful of amphibians and reptiles. It is important to realize, though, that there is a serious sampling bias. Testing for polarization sensitivity is difficult and so has usually only been looked for in migratory animals and those known to be especially good at navigation, such as certain desert ants. The true prevalence of polarization sensitivity is unknown.

The ability to sense the polarization of light has been divided into two types. One is known as "polarization sensitivity." Animals that have polarization sensitivity are not much different from humans wearing Polaroid sunglasses. Polarization affects the intensity of what they see—but without a lot of head-turning and thinking, they cannot reliably determine the angle or degree of polarization or even separate polarization from brightness. The other type is known as "polarization vision." Animals with polarization vision perceive the angle and degree of polarization as something separate from simple brightness differences. Behaviorally, this means that they can distinguish two lights with different degrees and/or angles of polarization regardless of their relative radiances and colors. This is much like the definition of color vision, which involves the ability to distinguish two lights of differing hue and/or saturation regardless of their relative radiances. While certain animals are known to distinguish differences in e-vector axis regardless of relative brightness and color, to my knowledge, nobody has varied degree of polarization at the same time (the number of parameters involved would make it a tedious experiment). So, while, technically, no animal has been shown to completely fulfill the requirements for true polarization vision, certain animals such as honeybees, the desert ant *Cataglyphis fortis*, cuttlefish, and the ever-amazing mantis shrimp are likely to have it.

While some have argued that the polarization of light may affect its transmission through the lens and cornea (via the reflection laws described earlier in the chapter), it looks like polarization sensitivity, whether rising to the level of polarization vision or not, primarily stems from the dichroic nature of the visual pigments' chromophores. As we discussed in chapter 4, the retinal chromophores found in visual pigments are long molecules. It turns out that they are most likely to absorb light if the long axis of the chromophore is parallel to the e-vector of the incident wave, and least likely to absorb light if the two axes are perpendicular. In other words, each visual pigment is a tiny Polaroid filter.

However, there needs to be structure at two higher levels of organization for this dichroism to provide polarization sensitivity to the animal: (1) the visual pigments must be at least partially aligned within the cell, and (2) the cells must be oriented within the eye. In invertebrates, this is accomplished by packing the pigments into the fingerlike projections (known as "microvilli") of the rhabdomeric photoreceptors and then arranging the cells within the eye so that the sets of projections point along different axes (figure 8.7). It is possible that, in some cases, the visual pigments are aligned by the cytoskeleton within each microvillus (Horváth and Varjú, 2004), but this is not a necessary condition. As we mentioned in chapter 4, visual pigments are membrane proteins and so are embedded in the surface of each microvillus. Even if they are arranged randomly on this surface, the chromophores will still have a weak statistical alignment parallel to the axis of the microvillus when viewed from above because the chromophores on each side of the microvillus are constrained by geometry to be within the plane of the membrane and thus parallel to the long axis of the microvillus. They will be tilted

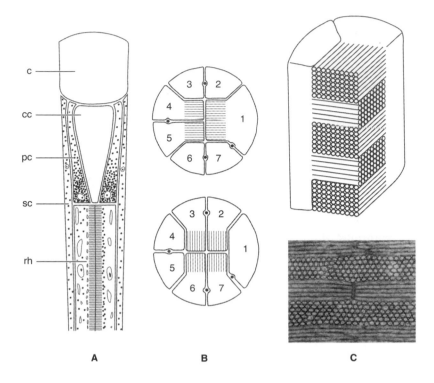

A   B   C

Figure 8.7: (A) A schematic longitudinal section through a generalized insect ommatidium, showing the corneal lens (*c*), the crystalline cone (*cc*), the primary pigment cells (*pc*), the secondary pigment cells (*sc*), and the rhabdom (*rh*). (B) The microvillar arrangements of seven photoreceptor cells in two consecutive layers of the rhabdoms. (C) (Upper) Neighboring retinula cells "dove-tail" their microvilli to create alternating layers of perpendicular microvilli. (Lower) Electron micrograph of microvilli in this arrangement. Images courtesy of Eric Warrant, Almut Kelber, Rikard Frederiksen, and Justin Marshall.

toward or away from the axis, but still be parallel as far as the dichroism goes. This can be tricky to understand, and the best way to prove it to yourself is to draw a set of short random lines on a skinny balloon and then look at it. You'll see that the lines in the middle of the balloon are random, but as you get to each edge, where the balloon curves away from you, the lines start to align. Some will be shorter than before (because you're seeing them foreshortened), but they still look parallel from this viewpoint, which is the same viewpoint the light sees. This statistical alignment is of course weaker than actually lining up the visual pigments deliberately, but still enough to provide polarization sensitivity.

How the photoreceptors and microvilli are arranged in the eye varies greatly by taxa. In many animals with compound eyes, you see an orthogonal array where half the cells within an ommatidium (facet) have microvilli that are perpendicular to those of the other half. In other ommatidia, you have cells with projections pointing along more than two axes. In still other invertebrate eyes, the projections within an ommatidium all point along the same axis, but this axis varies between neighboring ommatidia. The variety of arrangements is large and can be perplexing (see Waterman, 1981, and Wehner and Labhart, 2006, for extensive reviews), but all lead to the same result. Namely, even if the different photoreceptors all have the same visual pigment and view the same radiance, the amount of light they absorb differs depending on the orientation of the cells' microvilli relative to the dominant e-vector of the incident light. The animal then needs to compare the light absorption between these different photoreceptors to extract the degree and angle of polarization, just like our eye compares the amount of light absorbed by different neighboring cone cells to determine color.

Nearly all invertebrate photoreceptors have microvilli, and visual pigments are universally dichroic. In addition, it is common for photoreceptors to have some sort of regular geometric arrangement. For these three reasons, polarization sensitivity is nearly a default condition in invertebrates with rhabdomeric receptors. While this can be beneficial (as we'll discuss in the next section), it can also lead to trouble, particularly in animals with color vision, which involves comparing signals between neighboring cells containing different visual pigments. So, if the neighboring microvilli are oriented in different directions, the polarization of the incident light will affect the absorption in the photoreceptors and change the perceived color. Certain invertebrates (particularly butterflies for whom color vision is especially important) appear to get around this problem by twisting their rhabdoms along the axis of the light path (Wehner and Bernard, 1993). Thus, as the light passes through the cell and is absorbed, it interacts with microvilli at many different angles, which averages out the dichroism. Another solution would be to put a quarter-wave plate in front of the photoreceptors, thus converting any incoming linearly polarized light into circularly polarized light, which cannot be detected as polarized by photoreceptors. However, this only works if the axes of the quarter-wave plate are at 45° to the angle of polarization of the incident light. Regardless, I'm not sure if any animals actually do this, though one animal, which we will get to in a moment, uses a quarter-wave plate for a far more interesting trick.

As you can see from even the short discussion above, the neurobiology of polarization sensitivity and vision is surprisingly well-developed. I can't possibly do justice to it here, so as with bioluminescence, I am going to limit myself to a few special cases that highlight interesting optics.

The first "special" case is vertebrates. While polarization sensitivity is certainly rarer among vertebrates, it does exist. This has led to quite the controversy, because, unlike invertebrate photoreceptors, vertebrate cones and rods don't have microvilli that extend to the sides. Instead, the visual pigments are packed into the membranes of stacks of disks. Given this geometry, the incident light would "see" chromophores oriented at all angles. Frankly, no one has a universal solution to this and the mechanism of polarization sensitivity in vertebrates remains—along with the basis for animal magnetoreception—one of the two holy grails of sensory biology.

From what little we know (far fewer people work on polarization in vertebrates than invertebrates), some cells within the eye of polarization-sensitive vertebrates are polarization sensitive and some are not, which rules out the possibility that the optics of the eye are primarily responsible. In other words, it doesn't appear that salmon and trout are wearing polarized sunglasses, which leaves us to consider special adaptations in the photoreceptors themselves. There are three main hypotheses. First, it has been noted that certain cones in anchovies have their disks oriented along the long axis, instead of perpendicular to it, like typical cones. This roughly duplicates the geometry in invertebrate photoreceptors and so could provide polarization sensitivity. These cones are also backed by a reflecting tapetum oriented at an angle that would preferentially reflect polarized light back through the photoreceptor, increasing its sensitivity to a given e-vector. However, so far as I know, this system has only been seen in anchovies.

The second hypothesis involves twin cones, which are pairs of cones, usually containing the same visual pigment, that are stuck together, separated only by a long, flat membrane. The idea is that the morphology and refractive index gradient of the twin cones are such that one angle of polarization is transmitted more easily than the perpendicular one. However, measurements of the differential transmission of the cones show that the effect is

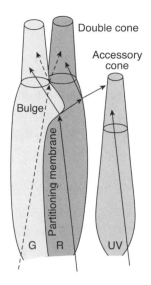

Figure 8.8: The double cone hypothesis for polarization sensitivity in the rainbow trout *Oncorhynchus mykiss*. The incident light enters from the bottom of the figure and, depending on its polarization, a fraction of it is reflected from the membrane between the red (R) and green (G) cone. The bulge in the membrane is angled such that the reflected light enters a nearby UV cone and is absorbed. Thus, the polarization of the incoming light affects the amount of light incident on and absorbed by the UV cone. Courtesy of Inigo Novales-Flamarique.

small, only about 1%–5%—probably not enough to account for the observed polarization sensitivity. A closely related hypothesis involves double cones (which usually have two different visual pigments). In this case, the polarization sensitivity is thought to be due to polarized reflections at the interface between the two cells. It is argued that the angle of the interface is such that incoming light is fairly well separated into its two polarization components, with one component reflected back to one cone and the other refracted into a third UV-sensitive cone (figure 8.8). Therefore the polarization of the incident light affect the amount of light incident on and thus absorbed by the UV cone (see Novales Flamarique et al., 1998, for details).

The final hypothesis notes that the photopigment-containing disks inside cones and rods are never truly perfectly aligned with the axis of the incoming light. This lack of alignment means that each disk will be slightly dichroic (due to statistical alignment from the tilt, much like the edges of the microvilli). Nick Roberts, in a technological tour de force, used optical tweezers to make individual vertebrate photoreceptors essentially stand on their heads on a microscope slide and then shot a tiny beam of polarized light down the length of the cell. This is a specialized application of a technique known as "microspectrophotometry," which is used to measure the absorbance spectra of photoreceptors. He found a small amount of dichroism, on the order of a few percent. Again, it is not certain if this is enough to explain the polarization-sensitivity of vertebrates. In addition, it is not known if the slight tilts of the disks occur in any sort of ordered fashion in the eye and whether the correct comparisons are made between cells with different tilts.

All in all, the basis of vertebrate polarization sensitivity is a contentious subject, often generating more heat than light. Sorting it out will be difficult, especially since few vertebrates exhibit a robust polarization response and multiple mechanisms likely exist, perhaps even within a single species.

Moving back to invertebrates, I would like to highlight a few species that do things differently, beginning with what took five years out of my life. I entered graduate school with no background in biology and even less in research. So despite the best efforts of my advisor, I chose a classic high-risk, low-yield subject for a thesis: polarization sensitivity in echinoderms, specifically brittle stars. In one way, I made a good choice. The calcite endoskeleton of echinoderms is a fascinating substance. It can take on many different ultrastructures, depending on taxa and function, but in general it is a complex spongelike network in which the remainder of the space is filled with organic material (cells, connective tissue, etc.) (figure 8.9). Despite this complexity, in almost cases an entire skeletal plate operates optically as a single crystal of calcite. As we discussed, calcite is highly birefringent. Therefore, if one of the two refractive indices matches that of the organic material, then light of one polarization sees a substance with a constant index and thus travels through it relatively unimpeded. Light of the other polarization, however, sees a substance with many interfaces and thus is scattered and attenuated. In other words, echinoderm tissue functions as a polarizer via differential scattering instead of differential absorption. This was actually one of the original ideas explored by Edwin Land when he was developing a commercial polarizer. I tested individual spines and plates from the brittle star *Ophioderma brevispinum* and found that certain classes of them did indeed polarize light. So far so good, but my continuation of the project, which involved proving that the animals responded behaviorally to polarized light, turned out to be nightmarishly difficult. I did eventually prove it (Johnsen, 1994), but would advise other people to approach behavioral experiments on echinoderms with care. The great invertebrate zoologist Libbie Hyman wasn't kidding when she said, "I also here salute the echinoderms as a noble group especially designed to puzzle the zoologist."

The second case involves polarization sensitivity via reflection from a tapetum. Marie Dacke (Dacke et al., 1999) found that the posterior–median eyes of the gnaphosid spider *Drassodes cupreus* (which point straight up at

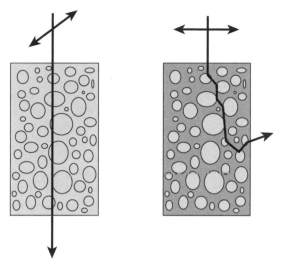

Figure 8.9: Schematic of the scattering polarizer found in echinoderm tissue. The tissue consists of birefringent calcite (background) and an organic matrix with an index that matches one of the two indices of the calcite (holes). (Left) Light of one polarization sees a material with few index changes and is transmitted well. (Right) Light with the perpendicular polarization sees a tissue with many index changes and is strongly scattered.

the sky), were highly modified. First, the eyes didn't appear to have any focusing apparatus, consisting only of a flat corneal window directly above the retina. However, below the retina was a bright blue tapetum that, due the usual laws of reflection, preferentially reflects light of one polarization (she discovered this by viewing the eye through a polarizer that she rotated, which made the reflection grow bright and dim) (figure 8.10). The polarization dependence of this reflection enhances the natural dichroism of the photoreceptors. The eyes are at 90° to each other, so comparing their responses gives the spiders excellent polarization sensitivity.

The final case is the last hurrah of the mantis shrimp (or the most recent, as my colleague Tom Cronin assures me). As we've discussed, these colorful and fluorescent animals have independently rotatable eyes with at least twelve photoreceptor classes and true polarization vision. If this didn't make them special enough, the animals have one final trick up their sleeve, a visual ability that so far is unknown in any other animal. Stomatopods can discriminate left-handed from right-handed circularly polarized light

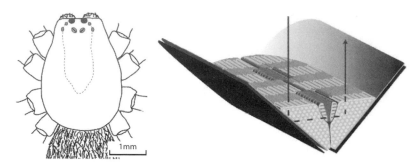

Figure 8.10: (Left) Top view of the prosoma of the gnaphosid spider *Drassodes cupreus* showing its various eyes. The posterior-median eyes (lower middle pair) have no ability to focus light, but are excellent polarization sensors because the usual array of microvilli is backed by a tilted mirror that preferentially reflects polarized light with certain e-vectors. (Right) Close-up of eye structure showing mirror and microvilli. From Dacke et al., 1999.

(Chiou et al., 2008). They do this by combining the normal invertebrate visual apparatus for detecting linearly polarized light with a quarter-wave plate. Any incident circularly polarized light gets converted to linearly polarized light, which can then be detected. The handedness of the circularly polarized light affects the e-vector axis of the converted linearly polarized light and so can be discriminated. Remarkably, the quarter-wave plate is actually a photoreceptor cell. In many compound eyes, each ommatidium contains seven large cells (usually called "R1–R7") lined up like a bundle of straws parallel to the axis of the incoming light and a much smaller eighth one (called "R8") sitting on top (see figure 8.7). Often, R8 is responsible for sensitivity to ultraviolet light, but in this case it has just the right thickness and birefringence to act as a quarter-wave plate. The two axes of birefringence are at 45° to the axes of the microvilli in R1–R7, further suggesting that this is not an accident but adaptive.

Even more remarkably, the quarter-wave plate works over a wide range of wavelengths. Because, by definition, a quarter-wave plate creates a 90° difference in phase between the two polarizations, you'd think it would only truly be a quarter-wave plate for one wavelength. This typical wavelength dependence of phase retardance is what causes the pretty colors you see when you look at birefringent materials between crossed polarizers. However, the R8 cell has managed to break what seems to be a simple and inviolable law of arithmetic (Roberts et al., 2009). It can do this because, as we discussed, there are two types of birefringence, intrinsic and form. In this case, the intrinsic birefringence is that of the membrane lipid molecules in R8, and the form birefringence is that of the parallel arrays of microvilli of the same cell. They turn out to depend on wavelength in opposite ways, and their sum (which is the measured birefringence) ends up being constant. Wow.

This remarkable set of adaptations and the only slightly less remarkable adaptations to sense the linear polarization of light in other animals beg the question, "Why on earth go to all this trouble?" It turns out that polarization sensitivity can be useful in a number of ways. Like the physiology of polarization sensitivity, though, the ecology of it has become a large and complex field. So the following section is only the briefest introduction.

ECOLOGICAL FUNCTIONS OF POLARIZATION SENSITIVITY

The ecological functions of polarization sensitivity can best be grouped by the source of the polarized light, with the main sources being: (1) polarized skylight, (2) polarized reflections, and (3) polarized underwater light. Functions related to polarized skylight are the best understood. They also all relate to some form of navigation or orientation, which makes sense because many animals use the sun's position for navigation, and even a hidden sun can be located using the polarization pattern of skylight. Ever since Karl von Frisch (1967) showed that honeybees could use the polarization of skylight for celestial navigation, a number of insects, crustaceans, and birds have been shown to do the same (reviewed by Horváth and Varjú, 2004). It is important to realize, though, that this is not as simple as it looks. Due to clouds, multiple scattering, and a host of other effects, the actual polarization pattern of the sky is more complicated than the theoretical models. Also, it's doubtful that insects are breaking out protractors and doing trigonometry in their heads to figure out where the sun is. Rudiger Wehner addressed this second issue by showing that bees, at least, mostly concentrate on the band of greatest polarization that is 90° from the sun and use a relatively simple algorithm to navigate. One thing has always bothered me, though. The sun is so intense that it's hard to understand why you would ever have trouble finding it. The only times it is truly impossible to find are when the sky is so overcast that there is no skylight polarization, anyway. This issue is usually swept under the rug of "It's always nice to have multiple sources of information," but I wonder. It takes a sky with a heavily overcast solar half and several large clear areas in the other half for polarization sensitivity to show its true worth as a sun finder. However, this morning had just such a sky, so maybe I'm overreacting.

A simpler navigational use of polarized skylight is as a guide for traveling in a straight line. This sounds trivial until you actually try it. Swim the back-stroke under an open sky or hike in a featureless terrain and you will soon find yourself going in circles. I have watched a lot of kids' swim meets and can tell you that the average backstroker swimming without lanes will turn 90° in about 15 yards. Few animals swim the backstroke, but many do need to go straight. A neat example can be found in dung beetles. They congregate on newly deposited dung piles and out of them form their balls. It is easier to steal a neighbor's dung ball than make your own (isn't it always?), so, once they have collected what they need, they race from the scene as quickly as possible. They begin this job at twilight, but continue on into the night. Marie Dacke discovered that their paths from the dung pile were straight when the moon was out and nearly random when the moon was gone (Dacke et al., 2003). As we discussed in chapter 3, the moonlit sky is nearly identical to the sunlit sky. This includes polarization, though in this case the polarization comes from scattered moonlight rather than scattered sunlight. Dacke and her colleagues held large polarizing filters over the dung piles. As they turned the filters, the animals turned as well, suggesting that they were using the polarization pattern of the moonlit sky, fairly amazing since it is at least one million times dimmer than the sunlit sky.

Another source of polarized light for which functions are somewhat understood are polarized reflections. Nature is full of shiny surfaces ranging from leaves to water to the iridescent surfaces of many animals, so polarized reflections are common. The two main functions related to this are water finding and signaling to conspecifics. Because the sun often hits water at intermediate angles, the reflected glare is usually polarized. This glare is annoying to humans and obstructs the view into the water, which is why fishermen's sunglasses are nearly always of the polarizing type. However, it can also be a way for an animal of little brain to identify water. This function has been explored in detail by Rudolf Schwind (e.g., Schwind, 1989) and is well established. It also has important consequences, because, while large polarized reflections in nature are nearly always from water, this selectivity vanishes when you include the human world. Glass, plastics, cars, asphalt roads, and even pools of oil all can have significant polarized reflections. Insects arrive at these surfaces thinking that they have reached water and then lay eggs or attempt to do some other appropriate task. In certain cases they never realize their mistake and guard the shiny roof of a Buick as their territorial pond. This has led Horváth et al. (2009) to coin the term "polarized light pollution." This may be a case of taking the "pollution" metaphor one step too far, but the article makes some good points.

The other major demonstrated use for polarized reflections is conspecific signaling. The first example of this was found by Nadav Shashar in the cuttlefish *Sepia officinalis* (Shashar et al., 1996). It has long been known that cuttlefish can change the color of their bodies for both camouflage and signaling. The range and dynamism of these patterns is amazing even for cephalopods. It has also been known that some of the patterns they produce are iridescent and thus based on coherent scattering. As we discussed in chapter 6, iridescence is closely related to specular reflection and so has the potential to be polarized. Nadav and his colleagues viewed the animals using a video camera fitted with a liquid crystal filter that acted as an electrically controlled polarizer. Over three successive video frames, the transmission axis of the filter went from 0° to 45° to 90°. Using this information, they were able to then reconstruct polarization video of the animal. They found that *S. officinalis* had stripes on its arms that could appear polarized or unpolarized at will. Since the change in polarization was accompanied by a change in iridescence, they tested to see whether the animal actually paid attention to the polarization change. They did this by having the animal view its own reflection through either a normal sheet of Pyrex glass or one that had many regions of birefringence due to repeated local heating. Animals viewing their reflection through the normal glass retreated when they saw their own polarization pattern (apparently cuttlefish cannot recognize themselves in mirrors) but usually stayed put when viewing their reflection through the treated glass that warped the polarization pattern. A later study by Alison Sweeney (the one who was nearly blown off the beach in chapter 3) used similar techniques to show that male *Heliconius* butterflies spent more time with females if the linear polarization signature from the females' iridescent wings was not altered via birefringent filters (Sweeney et al., 2003). Stomatopods, never to be outdone, appear to have at least three sources of polarized reflections: (1) red linear polarizers within the cuticle found in various locations depending on species, (2) blue linear polarizers found underneath the cuticle, often on the first maxillipeds, and (3) red circular polarizers found for instance on the keels of the telson of *Odontodactylus cultrifer*—the same species shown to have the ability to discriminate circularly polarized light (Chiou et al., 2005, 2008) (Plate 9). Tsyr-Huei Chiou has recently followed up on this work and shown that female mantis shrimp appear to find males less attractive when the polarization aspect of their appearance is removed (Chiou et al., 2011). As you can see from the dates on these papers, polarization signaling in animals is a new field. However, it looks to be a promising one.

We know the least about the ecological functions of the polarization of underwater light. As with the sky, the polarization pattern of the underwater light field contains information about the position of the sun and so could be used for navigation (Waterman, 2006). The connection between the pattern and the solar position is more complex, affected by waves, and vanishes with increasing depth, but one could easily see the advantages of a navigational cue in the relatively featureless pelagic environment. Either way, no one has so far shown that any aquatic animal uses the pattern to navigate (to be fair, though, no one has tried either). The only work on polarized light navigation under water involves animals so close to the surface that they are looking at the polarization pattern in the sky, distorted by the air-water interface (though, see Goddard and Forward, 1991, for an interesting exception).

Instead nearly all research on the biological use of underwater polarized light has focused on contrast enhancement, via either increasing visual range or breaking the camouflage of certain animals. As we discussed in chapter 5, underwater visual range is usually orders of magnitude less than it is on land, so optical engineers since at least the 1950s have tried to extend this range via various tricks. Early attempts centered around circularly polarized light, because reflection from extended surfaces (which you usually want to see) affects circular polarization differently than scattering from suspended particles (which get in the way). Researchers found that, by illuminating their subjects with circularly polarized light of one-handedness and then viewing through a filter of the other-handedness, they were able to get clearer underwater images (Mertens, 1970). However this requires bringing your own flashlight, which not all animals do (though it would be interesting to see if any bioluminescent searchlights are circularly polarized). Later attempts took advantage of the fact that the background light and path radiance (i.e., veiling light) were both polarized to remove them from the image. One such

Figure 8.11: Underwater image in raw form (left) and after using a polarization-based dehazing algorithm (right). From Schechner and Karpel, 2005.

process, developed by Yoav Schechner (Schechner et al., 2003) involves taking two images, one with a polarizer oriented such that the background polarized light is as bright as possible (usually when the transmission axis of the polarizer is horizontal) and one with the polarizer oriented so that the background light is as dim as possible. These images are then fed into a simple algorithm that removes much of the background and path light and essentially "dehazes" the image. When done well, the effect is remarkable (figure 8.11). It is not known whether any aquatic animals use their polarization sensitivity to do this, but it is certainly not inconceivable (see Johnsen et al., 2011). People talk a lot about biologists inspiring engineers, but this could be a case where engineers inspire biologists.

Among biologists, though, most hypotheses of contrast enhancement involve camouflage breaking in the water column. As we discussed in chapter

6, transparency and mirroring are two of the primary means of camouflage in the dangerously exposed pelagic habitat. While highly successful, they both have one potential flaw—they may affect the polarization of the light. Transparent tissues may depolarize the transmitted background light, or, if they are birefringent, alter the polarization from linear to elliptical. This latter possibility is especially problematic because, as we discussed, form birefringence is found in muscle, connective tissue, and other nutrient-rich substances. So your birefringent material, if visible, is more or less saying, "I'm good to eat." Mirrors have the related problem of potentially changing the polarization of the reflected light.

These are attractive hypotheses, especially since images of many transparent zooplankton between crossed polarizers show them to be strikingly birefringent. In addition, two more lab-based studies by Nadav Shashar and his colleagues showed that squid were better able to find birefringent clear beads than nonbirefringent ones and that cuttlefish preferred to attack silvery fish if the polarization of the light reflected from them was not removed (Shashar et al., 1998, 2000). However, as I mentioned in chapter 7, be wary of lab-based results and never fall in love with pretty pictures. My colleague Justin Marshall and I decided to see, in the field, what transparent and mirrored animals would look like to an animal with polarization vision and spent our spare time over a few oceanographic cruises and one trip to the Great Barrier Reef swimming around with various polarization imaging contraptions. We also collected animals to look at their birefringence in the lab. As we expected, the lab-based images showed impressive birefringence. However, none of this was visible in the field. While transparent animals did have an *in situ* polarization signature, it was mostly due to unpolarized reflections from the less transparent parts of the animals (Johnsen et al., 2011). Since these parts were fairly visible under normal vision anyway, the polarization information didn't make them more obvious. So what happened to the birefringence? The problem with the lab-based images is that they assume the animal is being viewed primarily via transmitted light. In other words, you put a bright light behind two crossed polarizers and view the animal between them. The underwater light field isn't like that. First of all, the light comes from all directions. Second, the down-welling light is far brighter than the horizontal background light. So, when we looked horizontally at our animals under water, we mostly saw them because they were scattering a small amount of the down-welling light toward us. The transmitted horizontal (and polarized) background light, which could have shown us the bire-fringence, was too dim in comparison. We also tried viewing the animals from below, but, as discussed, the down-welling light is barely polarized and so you won't see any birefringence. Thus, it looks like polarization sensitivity is not a good way to find transparent prey. For the silvery fish the story is less clear, but it seems that polarization sensitivity is also less useful than we ex-pected. At the moment, it looks like future research on polarization contrast enhancement should concentrate on whether animals use tricks similar to those engineers have developed for increasing visual range.

FINAL WORD OF WARNING

Unlike color and brightness, we can't see polarization and can't even imagine how it's perceived by animals. The physics of polarized light is subtle and there are more experimental pitfalls than average. In particular, experiments need extra controls because polarizers also affect the spectrum and intensity of transmitted light and because polarized light reflects differently off every surface of our setup depending on angle, leading us to possibly mistake an animal's light sensitivity for polarization sensitivity. So, while buying a polar izer is easy, using it correctly in a biological setting is not. I normally don't like to warn people or appeal to authority since I think it hurts creativity, but this is one case where it is best to start by collaborating with an established expert.

FURTHER READING

The only book on polarized light written explicitly for biologists is *Polarized Light in Animal Vision: Polarization Patterns in Nature* by Gábor Horváth and Deszö Varjú. Luckily it's a good one and a fairly comprehensive review of the field. Aside from this, you are left with review articles and book chap-ters (those by Tom Cronin, Justin Marshall, Talbot Waterman, and Rudiger Wehner are especially good). Talbot Waterman's 1981 chapter in *The Hand-book of Sensory Physiology* stands out as the review article we would all love to write. Almost two hundred pages long with beautiful figures, it covers just about everything that was known at the time and is still worth reading as an introduction to the field (or to look up a long-forgotten fact). Actually, nearly every chapter in the multivolume series is a masterpiece and worth reading.

Polarized Light in Nature by Gunther Können covers all the different sources of polarized light, some of which—like lava—you likely have never considered. Lynch and Livingston's *Color and Light in Nature* (first mentioned in chapter 3) also discusses natural sources of polarized light.

The early chapters of Craig Bohren's *What Light Through Yonder Window Breaks?* contain a see-it-for-yourself discussion of polarized light and birefringence. His and Clothiaux's *Fundamentals of Atmospheric Radiation* has a more technical, but still accessible chapter on the subject. Read it, even if only to learn just how wrong most textbooks are.

The best short introduction to the field is William Shurcliff's *Polarized Light: Production and Use*. It has some math, but the text is so good that you'll hardly notice. Shurcliff worked for Polaroid Corporation during the glory days, when the first sheet polarizers were being invented. You'll be amazed at how many varieties there are.

Jearl Walker's *Flying Circus of Physics* has a number of demonstrations and puzzles that help explain the phenomenology of polarized light. Walker wrote the Amateur Scientist column for *Scientific American* for many years and does a wonderful job of making the complex both simple and fun.

If you're interested in the early history of animal navigation and polarization sensitivity in animals, read Karl von Frisch's *The Dance Language and Orientation of Bees*, which summarizes fifty years of his work up to 1967. It is beautifully written and a stunning example of just how far careful observation, experiment, and common sense can get you.

CHAPTER NINE

Measuring Light

I really hate this damned machine, I really ought to sell it. It
never does what I want, but only what I tell it.
—AUTHOR UNKNOWN (A programmer's lament, from Dennie
L. van Tassel, *The Compleat Computer*)

While it can often appear as simple as turning on a machine and recording a
number, light measurement has many nasty pitfalls. It is easy to measure the
wrong thing, to measure the right thing in the wrong way, or to have a worthless number and not know it. The main reason for this is that we don't measure light in our daily lives. Since childhood, we develop an intuitive sense of
weights, lengths, area, temperature, and so on. For example, we can guess
someone's height to within 5% and weight to within 10%–20%. However,
even after a decade of measuring light, I cannot not tell you how bright my
office is on this overcast morning to within even an order of magnitude.

Part of the problem also lies in our miraculous eyes, which can see in
environments that range over ten orders of magnitude in irradiance, and in
our brains, which make this range seem far less than it is. In order to do
this, our visual system employs a logarithmic response that is continually
adapting to the current irradiance. I often tell my students that, if they were
to measure the radiances of all the pieces of white paper they see over a
typical day, they would range over a couple of orders of magnitude and vary
in color from pale blue to deep orange. However, our eyes and brains soften
these differences, so we always see the paper as white. This remarkable ability allows us to perceive the world as a relatively constant place, but wreaks
havoc with our ability to ground-truth our light measurements. While we
would instantly be suspicious if our scale told us that a rabbit weighed three
grams, a similar mismatch between a light measurement and reality would
easily go unnoticed. So we are left trusting the machines, which leads to
problems.

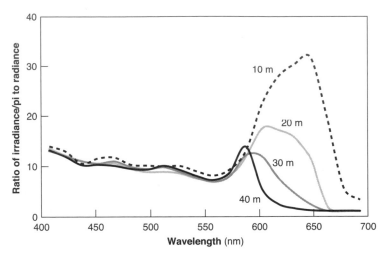

Figure 9.1: Ratio of side-welling irradiance to horizontal radiance as a function of depth and wavelength in clear oceanic water. Note that the relationship depends strongly on both parameters. (Irradiance is divided by π so that its units match radiance.)

By far the biggest problem is not measuring what you think you are measuring. I have reviewed many manuscripts in which the authors took thousands of light measurements, but all the wrong kind for the questions they were asking. In some cases, the measurements could be converted to the correct ones, but in most cases they couldn't. Most of these involved measuring radiance instead of irradiance (or vice versa) or using integrated units, such as lux.

The confusion between radiance and irradiance is especially depressing because, while it is easy and often cheap to convert a detection system from one to the other beforehand, it is impossible to convert the measurements after the fact. Because irradiance measurements collect light from an entire hemisphere, they generally bear no relationship to radiance measurements that collect light from only one small region. Even under water, where the light field appears highly uniform, the relationship between radiance and irradiance depends on viewing direction, depth, and wavelength (figure 9.1).

Similarly, a light measurement done in photometric units cannot be converted to other units unless one knows the spectrum of the light. As mentioned in chapter 2, photometric measurements are weighted integrals over the entire visible spectrum. Once this integration is done (generally by a colored filter that matches human visual response), it is impossible to recover the original spectrum. For example, a 555 nm green laser has one hundred times the luminance (perceived brightness) of a 425 nm blue laser that emits the same number of photons, because a photometric detector is more sensitive to green light. In fact, one can't even convert photometric measurements into other integrated measurements, such as total number of photons or watts integrated over the visual range.

Other integrated measurements are no better. Botanists and oceanographers often measure what they call photosynthetically active radiation (PAR), which is the total number of photons between 400 nm and 700 nm. Again, because underlying spectra are unknown, the voluminous databases of PAR cannot be converted into any other measurements.

Conversions between different integrated measurements such as lux and PAR can be done, of course, if one knows the spectrum of the light. However, in most cases, if one knows the spectrum, there is no need to take integrated measurements, since one can easily integrate the data later. Indeed, although they can't convert radiance to irradiance, spectra are a central clearinghouse that can be converted into almost every other light measurement. In fact, the benefits of spectral measurements so far outweigh the costs that the best purchase for anyone measuring light is a spectroradiometer. While these instruments used to be large, expensive, and finicky, you can now buy a rugged spectrometer no bigger than a computer mouse for the price of a laptop. There are far more expensive models that work in low light or have extremely high spectral resolution, but these are not needed by most biologists. The rest of this chapter explores the use of spectrometers in light measurement, beginning with equipment issues and then going on to specific applications.

Types of Spectrometers

The first issue is buying the right thing. Again, names get in the way. For most biologists, a spectrometer is a heavy box that never moves from its spot on the counter in a molecular biology lab. These devices, also called "spectrophotometers" (despite having no relationship with photometric units), measure the absorbance spectra of solutions in cuvettes. They are useless for any other sort of spectral measurement. Far more useful is a spectroradiometer,

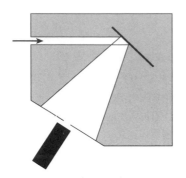 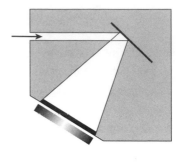

Figure 9.2: (Left) Scanning spectrometer. Light enters through slit at upper left, is spread into a rainbow by a diffraction grating. One small part of this rainbow exits the spectrometer through another small slit and is detected. Which part gets through is determined by rotating the grating. (Right) Multichannel spectrometer. As before, light enters through a slit and is spread into a rainbow by a diffraction grating (that is now fixed in place). The entire rainbow shines on a linear array of detectors.

also called a "spectrometer." The central difference is that a spectrophotometer has an interior light path, while a spectroradiometer has a detector that faces the outside world. Spectroradiometers are also generally much smaller and cheaper. To save on syllables, I will refer to spectroradiometers as spectrometers from here on.

Spectrometers come in two flavors, scanning and multichannel. While all spectrometers take incoming light and split it into a rainbow (almost always using a diffraction grating), the two types differ on how the light from this rainbow is detected (figure 9.2). Scanning spectrometers shine the rainbow on a black wall that has a small slit. Only the part of the rainbow that passes through the slit is detected. The whole spectrum is recorded by turning the grating, which rotates the rainbow and thus the spectral region that passes through the slit. So the spectrum is scanned over a period of time.

There are some advantages to this. First, the spectrometer only needs one detector, which can be large and thus more sensitive. Second, because the slit masks almost all the extraneous light, scanning spectrometers minimize what is called "stray light." Even though the insides of spectrometers are painted black and have other precautions, light of one color can bounce around and end up in the wrong place. In particular it can end up being recorded as light of a different wavelength. Even small amounts of stray light can ruin measurements of dim wavelengths. This especially wreaks havoc with measurements of UV radiation in the presence of the much brighter visible light. So full-spectrum measurements of solar radiation are often done with scanning spectrometers.

Finally, and most importantly, scanning spectrometers can change their gain during a scan. Many natural irradiance spectra can vary by orders of magnitude over their wavelength range. On a sunny day, for example, the down-welling irradiance at 300 nm can be a hundred to a thousand times dimmer than that at 450 nm. So the gain that works well at 450 nm may make the 300 nm measurement indistinguishable from the noise. Similarly, a good gain for 300 nm will almost certainly saturate the detector at 450 nm, or even destroy it. Variable gain solves this, resulting in noise-free spectra over large intensity ranges.

Scanning spectrometers also have significant problems, perhaps the worst of which is that they are slow. The scans take at least a few seconds and, at low light intensities, can take up to twenty minutes. I once spent hours in a submarine at 500 meters depth, waiting for a scanning spectrometer to take its measurements. This is tedious, but more importantly can lead to distorted spectra if the light levels or optical properties of your subject change during the scan. On the same cruise, we turned off all the lights on the ship to take a light pollution–free irradiance spectrum of a moonless night. The scanning of the spectrum was almost done, when the captain ran up with a flashlight to tell us that the Coast Guard was coming fast, assuming that our darkened ship was running drugs. The spectrum looked good up to 600 nm (which is when he showed up with the flashlight) and then shot through the roof.

Scanning spectrometers also tend to be larger and more expensive, both by at least a factor of ten. So, unless you need high sensitivity or low stray light, multichannel spectrometers are a better choice.

Multichannel instruments do not have a slit in front of the detector, but instead shine the entire spectrum on a line of detectors, usually a CCD array or row of photodiodes. This means that the entire spectrum is recorded at once, over a period as short as 3 milliseconds. As you might guess, the advantages and disadvantages are the opposite of those for a scanning system. Multichannel systems are usually less sensitive, because the detectors are smaller. However, one can average many spectra to improve sensitivity. This can make them as slow as a scanning system, but because each spectrum is collected over a short period, changes in light level during a long set of averaged scans do not distort the spectrum as badly. A bigger problem is that multichannel spectrometers have a constant gain for the whole spectrum. This, combined with much greater stray light, means they have a smaller dynamic range, generally between 250:1 and 1000:1 under ideal conditions. So if the intensity of the brightest wavelength is many times the intensity of the dimmest wavelength, you will not be able to measure both well. A low gain will give you a noisy signal at the dim wavelengths; a high gain will saturate the brighter wavelengths. While not too much of a problem for indoor work, for outdoor measurements this can be a real pain. Despite this, multichannel instruments are generally the better buy for a biologist.

There are two hybrid systems that combine some of the advantages of scanning and multichannel spectrometers. The first are multiband radiometers. While not spectrometers per se, they function as crude versions of them by having five to ten spectrally selective channels, generally 10–20 nm wide, spaced over the wavelength range. These channels have their own variable gains and also filters that make them extremely selective. Systems of this sort are often used for measuring UV radiation, because naturally occurring UV is so much dimmer than visible light, and UVB is so much dimmer than UVA. One such system, the PUV-500 by Biospherical, has six 10 nm wide channels centered at 305 nm, 313 nm, 320 nm, 340 nm, 380 nm, and 395 nm in addition to a PAR sensor. Aside from the obvious loss of spectral resolution, the other big disadvantage is that multiband spectrometers tend to cost as much as scanning spectrometers. The other instrument, known as an "optical multichannel analyzer" (OMA), is a multichannel spectrometer with an intensification system in front of the detector array. This makes them as sensitive as scanning systems, but without the problem of spectral distortion. While wonderful for measurements of variable and dim light such as bioluminescence, they are extremely expensive, roughly $100,000.

While scanning, multiband, and intensified spectrometers are necessary in certain cases, they are usually not the best purchase, especially for a first spectrometer or if money and portability are serious considerations. Of course, light-measurement needs in biology can be as diverse as the field itself. However, in my experience, multichannel spectrometers are the best choice for over 90% of applications, and the rest of this chapter pertains to them.

How to Use a Spectrometer

All spectrometers, scanning or multichannel, are essentially cameras that take very dull pictures. As such, they primarily care about two things: aperture and shutter speed. Together these determine how much light enters the spectrometer during a measurement. Each also has other less obvious effects on the quality of the spectrum.

Aperture is almost always controlled via hardware. In other words, you change it with your own hands. Nearly all recently made multichannel spectrometers are designed to work with fiber optic cables. While they generally cost a few hundred dollars each, these fibers are quite convenient. They allow you to measure light in all sorts of narrow spaces and many meters from your spectrometer (useful if you want to measure light under water). They also protect you from the light, which is important for UV work. For these systems, the aperture is just the cross-sectional area of the fiber. If you have multiple fibers in series, the aperture is more or less that of the smallest fiber. Fiber diameters generally range from 10 μm to 1000 μm, so just by changing fibers you may be able to change the aperture by 10,000 fold (remember that aperture is an area and so is proportional to the square of fiber diameter). I say "may," because most spectrometers also have an entrance slit, which complicates things. If the entrance slit is narrower than the fibers you are using, then changing fibers has only a linear effect on the amount of light that enters the spectrometer, not a square effect. For example, if a spectrometer has a slit that is 100 μm wide and several mm long, then going from a 500 μm to a 1000 μm fiber comes closer to doubling the amount of light, rather than quadrupling it. This sounds nitpicky, but becomes important when you are miles out to sea or on top of a mountain with those two fibers only to learn that they don't give you the sensitivity range you expected.

Aside from controlling how much light enters the spectrometer, aperture also affects the resolution of the spectrum in multichannel spectrometers, with narrower apertures improving spectral resolution. As mentioned above, these spectrometers send the incoming light to a grating that then shines a rainbow on the detector array. Even if the spectrometer is fitted with a quite small entrance slit, the cross-sectional area of the beam entering the spectrometer affects the size of the spot that hits the grating and thus the resolu-

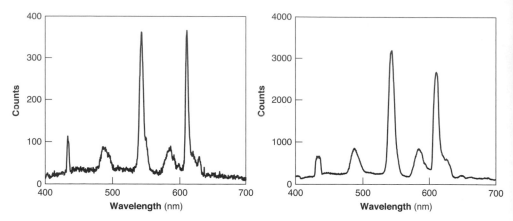

Figure 9.3: Uncalibrated spectra of room light (mixture of fluorescent light and daylight) taken using 100 mm (left) and 1000 mm (right) diameter fibers. Note that the spectrum using the larger diameter fiber is higher (in counts), smoother, and has wider peaks. The spectrum is smoother due to an increase in the signal to noise ratio, but the wider peaks are due to a loss of resolution that comes from using larger diameter fibers. Also note that, even though the larger fiber has one hundred times the area of the smaller, it only has about ten times the signal, for the reasons described in the text.

tion of the spectrum (figure 9.3). While this doesn't matter much for natural spectra, which are quite broad, it does matter if you are trying to characterize a fluorescent bulb or an iridescent butterfly, both of which can have narrow spectral peaks. Aperture also tends to affect stray light, because a larger beam of light inside the spectrometer is more likely to bounce where it shouldn't. As a general rule of thumb, 400 μm fibers attached to spectrometers with 200 μm slits give spectra with enough detail for almost all biological work. I only use narrower fibers to prevent saturation of the spectrometer when working outdoors.

The other main parameter for light collection—shutter speed—is always controlled by software. I suppose there may be a spectrometer somewhere with a dial for this, but I have never seen one. In most spectrometer software, shutter speed is referred to as integration time and is measured in milliseconds. Unlike with a camera, there usually isn't an actual shutter inside the spectrometer. Instead, integration time controls how long light is allowed to interact with the detector before the detector's state is measured by the computer. What happens inside a detector depends on its type, but I tend to vi-

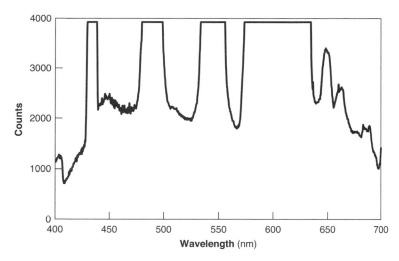

Figure 9.4: Uncalibrated spectrum of room light showing saturation in four spectral regions. Note the flat tops where the signal can't get any higher.

sualize them all as a row of jars being filled by a rain of gumballs. Integration time is how long you wait before emptying the jar and counting your candy. Increasing integration time, just like increasing shutter speed in a camera, makes the spectrometer more sensitive to light.

Unlike aperture, changing integration time does not affect spectral resolution, but the range over which you can adjust it is limited. At the lower end, most spectrometers will not accept integration times below a few milliseconds. This seems short, but can still be too long when working outdoors on a sunny day. At the high end, the spectrometer will saturate, which means that any further increases in light cannot be measured (figure 9.4). Essentially, the jar is overflowing. The saturation point generally depends on what is known as the "well depth" of the detector and on how the data is converted to digital form for your computer.

As mentioned above, a central problem with multichannel spectrometers is that they have one integration time for the whole spectrum. So, more often than you might guess, you have to saturate one part of the spectrum in order to clearly see another part. This won't harm the spectrometer, unless you have a fancy spectrometer that uses photomultiplier tubes as detectors. However it increases stray light and can sometimes make the software act strangely. If you need to see a portion of a spectrum that is much dimmer than the rest, it is usually best to filter out the brighter portion of the spectrum before it enters the spectrometer.

So, what if the light you want to measure is dim at all wavelengths? Can you increase the integration time to anything you want? Unfortunately you cannot—because of noise, which brings us to the next section.

GETTING THE MOST OUT OF YOUR SPECTROMETER

Some spectra look clean, some look ragged. Aesthetics aside, ragged spectra can hide useful data and, even worse, create spurious data such as false peaks. Unless one is working at the absolute limits of the spectrometer's sensitivity, which is rare, it is possible to greatly improve the appearance and usefulness of your spectra. The biggest issue is noise, or rather, the ratio of the signal to the noise. The lower the ratio, the worse things look. Once the ratio gets below about five, spectra start to look bad indeed.

In modern spectrometers, the main source of noise is not noise at all, but proof of the quantal nature of light's interaction with matter. As we discussed in chapter 1, light has many wave characteristics, but is detected in discrete units. So, for dim light or short collection times, the detectors in a spectrometer are much like the ground at the beginning of a rainstorm, with detection following the same random distribution as raindrops. Random events of this sort follow a Poisson distribution, which means that the standard deviation equals the square root of the average. So the signal to noise ratio, which is the quotient of the average and the standard deviation, is also equal to the square root of the signal ($N/\sqrt{N} = \sqrt{N}$).

Although we think of photons as being impossibly numerous, the number falling on a tiny detector over a period of a few milliseconds is not large. For example, the irradiance on a cloudless beach at sunset is about 10^{10} photons/cm²/s/nm. A spectrometer fitted with a 200 μm diameter fiber has a detection area of 0.00031 cm² ($= \pi \times 0.01^2$). Over a collection time of 3 ms this fiber would collect about 10,000 photons/nm, or about 3,000 photons/channel for the 0.3 nm resolution spectrometer I normally use. However, the cosine correctors used to measure irradiance (see chapter 2) are inefficient. Mine has an efficiency of about 10%, which reduces the number of photons to 300. This gives a signal to noise ratio of about 17. This is a good number,

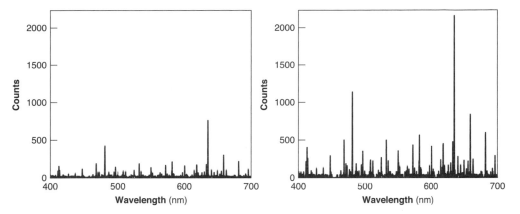

Figure 9.5: Dark noise for integration times of three seconds (left) and nine seconds (right). Dark noise is roughly proportional to integration time, and certain pixels on the detector always have higher dark noise values than others.

but a cloudless beach at sunset is quite bright. Many measurements in biology involve far lower light levels.

Unfortunately, one can't simply increase the integration time without limits, due to a second source of noise called "dark noise" or "dark current." The detectors in modern spectrometers are extremely reliable. However, they will occasionally measure light that is not actually there, mostly because thermal noise triggers a false event in the detector. While dark noise can safely be ignored for integration values under one-tenth of a second, at higher values it becomes a nuisance (figure 9.5). For 5-second integration times, some of the channels will have dark noise equal to one-quarter of the total saturation value. For 10-second integrations, many of the channels are over a quarter of the saturation value and some are over a half. The one good thing about dark noise is that it is pixel specific. If the channel of your spectrometer that records 600 nm light has a dark noise of 400 counts over 5 seconds today, it will have roughly the same counts of noise tomorrow (unless the temperature changes). Some pixels have low dark noise, and some pixels (often known as "hot") have especially high dark noise. For these reasons, all spectrometer software has a process by which you subtract this noise from the signal. The details vary, but essentially you put the detector end of your spectrometer in a dark place and take a spectrum. This spectrum is then subtracted from all future spectra. The amount of dark noise depends nonlinearly on integration time, so you need to take a new dark spectrum whenever you change it. In fact it's a good idea to take a new dark spectrum when you change any acquisition parameters. It also depends nonlinearly on detector temperature, which can vary, so it's important to take dark spectra at the same time you take your actual data.

This works wonderfully until you have integration times so long that your measurement plus the dark noise saturates the detector. Suppose your detector saturates at 4000 counts and that the dark noise in your 600 nm channel is 3000 counts for a ten-second integration period. You attempt to measure a light that should have 2000 counts at 600 nm. This gives 5000 counts, but you have saturated the detector, so it reads only 4000 and then subtracts 3000 for the dark noise, leaving you with the wrong measurement of 1000 counts. In reality, even before you get to saturation, variations in dark noise will make the spectrum messy. On my spectrometer, the integration limit seems to be about 10 to 20 seconds.

Since the dark noise is primarily thermal in origin, cooling the detector array reduces it. However, this causes other problems like condensation and higher power consumption (which can be a problem in the field). Remember also that thermal noise depends on degrees Kelvin, so cooling the spectrometer from room temperature to freezing drops the relevant temperature by only 7%.

So what to do? Fortunately, there are two solutions, both of which involve averaging. Which one you use depends on whether you are willing to sacrifice spectral or temporal resolution. The easiest solution is to average neighboring channels. As mentioned above, most spectrometers have better resolution than most biologists need, usually at least three channels per nanometer. The natural spectral world is usually far smoother than this. In fact, studies on the statistical properties of natural spectra (of rocks, trees, animals, etc.) have found that it only takes three to four principle component spectra to account for over 95% of the variation (Chiao et al., 2000). This means that, with the right filters over four detectors, you could re-create almost any natural spectrum from just the measurements in these detectors. It is thought that this is why most animals have no more than four visual pigments. So dropping your resolution from 0.3 nm to 2 nm is no big deal.

It is possible to average neighboring channels while collecting the data. Most versions of spectrometer software refer to this as "boxcar-ing." All you do is input the number of channels you want to average. However, I would strongly suggest that you not do this. You can always average neighboring channels later in a spreadsheet program, but if you do it while collecting the data you will never be able to recover the original values. Maybe you don't think you need the original data, but life has a way of surprising you.

The other solution is to average multiple spectra over time. This allows you to have as large an integration time as you want without worrying about saturating the detector or getting appalling dark noise. Of course, this won't help you if the event you are measuring is short, but for stable sources, spectral averaging is quite useful. However, there is also a final source of noise that limits what you can do. This noise, called "read-out noise," happens when the signal is transferred from the detector to the computer. Due to this noise, averaging even large numbers of spectra won't help much if you have a spectrum with especially low values (say 1% –5% of detector saturation). In other words, an average of one hundred 10 ms exposures will not necessarily give you the same signal to noise as an average of ten 100 ms exposures, even though both have a total integration time of one second. I only learned this several years ago after many painful phone conversations with an engineer at the company where I bought one of my spectrometers. Therefore, before you even think of averaging, you need to get enough light into the spectrometer to have something useful to average. The basic rule of thumb is to increase integration time and fiber diameter as much as you can and then, if you still need it, average spectra. As with boxcar-ing, it is best to save all your spectra and to average later in a spreadsheet program. Use reason, though. If you are averaging dozens of spectra, let the spectrometer software do it.

One final warning about averaging: whether you average over channels or time, remember that your signal to noise ratio (i.e., the tidiness of your spectrum) is proportional to only the square root of the number of averages. Therefore boxcar-ing over nine channels only increases your signal to noise by a factor of three, and averaging 100 spectra only increases your signal to noise by a factor of ten. So, while it is possible to convert what appears to be total garbage into a clean spectrum by averaging, the cost may be too high. My rule of thumb is that, no matter what you do, commercial multichannel spectrometers usually stop working at about the same light levels at which we switch from color to black-and-white vision. So if you want to measure nocturnal landscapes, bioluminescence, or other dark events, you will need to spend about ten times the money and get a fancier machine.

The final issue to consider is stability. With the exception of irradiance, optical measurements are highly sensitive to geometry. In other words, if your light source, detector, or target move even slightly, your measurement may change by an order of magnitude. Before you collect spectra for real, you should prove this to yourself. Plug a fiber optic cable into your spectrometer and point it at a flashlight beam. Move and turn the end of the cable and watch what happens to the spectrum. If you are like me, you will be depressed by how much effect even small movements have. Because eyes are logarithmic, we often only need to know absolute measurements to within an order of magnitude. However, many important measurements, like transmission, reflectance, and scattering are normalized by a reference measurement. If your optical setup moves between your reference measurement and your test measurements, the latter will be worthless.

So, everything needs to be mounted securely. While you can tape things down in an emergency, you will be surprised at how poorly this works. To get any real repeatability, you need to hold things in clamps that are mounted to a sheet of metal. Optics companies make sheets with predrilled holes called "breadboards" that are indispensable for doing optical measurements. You can buy giant optical tables for over $10,000, but a $1' \times 2'$ sheet for $100 works for most things. With one of these and a small set of clamps, screws, and rods, you can set up almost any measurement. It does require ingenuity, though. I usually tell people that an optical measurement is 99% setup and 1% measurement. It helps if you grew up with an erector set. Not surprisingly, optical mounts are often referred to as Tinkertoys.

The power output of the light source needs to be as stable as the physical setup. There is no sense in building the Fort Knox of optical rigs and then powering it via a flashlight with a fading battery. Again, the logarithmic nature of our vision hides the true variability of many light sources. Incandescent lights slowly brighten over the first fifteen minutes as the whole lamp reaches an equilibrium temperature. Both battery and outlet-powered lights are subject to the vagaries of their power sources, and certain light sources, such as arc lamps, are inherently unstable.

The best solution is to buy a light source that is designed for optical measurements and has a stabilized power supply. The second-best solution, which works fairly well, is to check the stability of your source often (every five to ten minutes) and then take a new reference measurement if it has drifted.

Finally, unless you are working in a pitch-black room, the environmental light levels may change and affect your measurements. Working in the dark is no fun, so it is best to design a setup that minimizes the influence of room light, either by covering the setup during measurement or by ensuring that the test light is by far the dominating factor at the detector. However, the sun can pop out from behind a cloud and affect even the most careful plan. A student of mine was measuring the reflectance of hundreds of blue crab claws in a crabber's shed at the beach and sadly discovered later that the black cloth she used to block the sun wasn't as black as she had hoped. The data were unsalvageable.

So, on to the actual measurements. In my own ordering of increasing difficulty, they are reflectance, transmission, irradiance, radiance, and scattering. Reflectance, transmission, and scattering measurements are normalized by simple standards and do not require absolute calibration of the spectrometer. Absolute irradiance and radiance measurements require absolute calibration, which makes them a bit more challenging. Scattering measurements are the most difficult due a number of geometric and sensitivity issues that I will discuss in that section.

REFLECTANCE

A nice thing about reflectance and transmittance is that they are unitless, which means you don't have to worry about calibrating your spectrometer. However, this advantage is often outweighed by geometric issues. Reflectance is particularly scary in this way.

There are a few different subtypes of reflectance, but the one most useful to biologists is the radiance of an illuminated object normalized by the radiance of a perfect diffuse reflector that is illuminated by the same light. This simple description overlooks the fact that this ratio depends on both where the incident light comes from and where you put your fiber optic detector. So, to completely describe the reflectance of a material, you must measure it for every angle of incidence and every angle of detection, each over the whole hemisphere of possibilities. Even ignoring the wavelength dependence (which is always there), you end up with a four-dimensional data monster known as the "bidirectional reflectance function" (BRDF). As you might guess, it takes a long time to measure a BRDF and even lon-

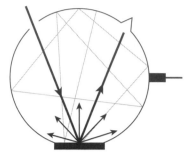

Figure 9.6: Diffuse and specular reflectance being measured in an integrating sphere. The sample (on the bottom) is lit by a beam of light (heavy arrow). Some of this light reflects in a mirrorlike fashion and is caught by a light trap (upper right). The rest is reflected in all directions (gray lines), some of which is collected by the fiber optic cable on the right. In some integrating spheres, the light trap can be replaced by a white surface (via a small lever). If this is done, the sum of both specular and diffuse reflection is measured.

ger to understand what you have. Unless you have a desperate need for this level of information (and work on something that is flat and unvarying), leave this sort of work to the people who have the money to do it and can actually use the data—computer animators like Pixar, and remote-sensing satellite designers like NASA.

Luckily, the reflectance of most biological objects (structural colors aside) can be reasonably well approximated as a sum of diffuse and specular (mirrorlike) reflectance (figure 9.6). This means that a fraction of the light will be fairly evenly reflected over all directions and the rest will bounce off the surface like a mirror, following the classic "angle of incidence equals angle of reflection" rule. I usually picture a fuzzy ball of light with one spike. However, the ratio of diffuse to specular reflectance in biological materials can vary from zero (the mirrors in scallops eyes) to one (dusty camels).

The most straightforward way to measure reflectance is to shine a light on your surface and use your detector to measure how much light bounces off it. The standard setup shines the light straight down onto the surface and measures the reflected light at 45°. This way, the detector doesn't see any of the specularly reflected light from the source.

The light can be as simple as a flashlight, but needs to have enough power in all the wavelengths you care about. For example, you cannot measure ultraviolet reflectance without some ultraviolet radiation coming out of your light source. The computer will happily give you a number, but it will be garbage.

The light must also evenly illuminate the appropriate region, which may require a diffuser. Collimating the light by putting a convex lens exactly one focal length from the source also works well. The detector must view only the illuminated region. Any unlit regions in the field of view will lower the measured reflectance (since the probe will be looking at dark regions). Since your detector is usually a fiber optic cable, a convenient way to determine its field of view is to shine light through it, simply attaching a lamp where the spectrometer usually goes. The area illuminated is also the spectrometer's field of view. This simple trick, based on the fact that light doesn't care which way it is going, has led to the design of reflection probes. These probes are parallel bundles of fibers; some illuminate the surface, the rest view it. They are convenient, but, because moving them changes both the illumination distance and the viewing distance, small changes in placement can have a big effect on the measurement. Use them only if they are fixed in place.

If you have $1,500, and the surface you want to measure is reasonably flat and at least one cm across, use an integrating sphere with a built-in light source (figure 9.6). This clever gadget shines light onto your surface and then collects the reflected light in all directions, a portion of which is detected by a fiber optic cable mounted inside the sphere. Mine has a nice addition, a lever that opens and shuts a light trap that removes all the specularly reflected light. This way you can first measure all the reflected light, and then just the diffusely reflected light. The difference between the two is the specularly reflected light.

Of course, all the detector records is a light level. To calculate reflectance, you need to normalize it by the radiance of an object that is a perfect diffuse reflector. This standard reflects all the light that strikes it, and its radiance from any viewing direction is always the irradiance of the incident light divided by π. Most people use a foam version of Teflon with microscopic pores called "Spectralon." It's a proprietary material and not cheap, about $300 for the usual one inch diameter standard, but is indispensable and lasts forever. To use it, you record the light that bounces off your standard and then click a button in your software that tells the computer this light level means 100% reflectance. Then you are ready to go.

If your light source has a stable output and it and the detector don't move around, reflectance measurements are about the easiest optical measurements to do. However, there are a couple potential pitfalls. First, many things are not as opaque as you might think. For example, I often have people measuring the reflectance of flowers in my lab. While flowers appear opaque, many actually transmit a fair bit of light. Therefore, the measured reflectance depends on what is behind the flower. If you put the flower on a white table, you are measuring the sum of two things: (1) the light reflected from the flower, and (2) the light transmitted through the flower, reflected by the table, and then transmitted through the flower again. One way around this is to put the flower over a hole in an otherwise closed black box. Then the light that is transmitted through the flower mostly gets swallowed up by the box. A more portable solution is to put the flower on something black. The average black piece of paper won't do, because its reflectance is usually at least 5%. I often use a visible-block filter, as they have a high concentration of compounds that strongly absorb visible light. As long as you are careful to avoid specular reflection, these filters are great portable light traps. Most paper absorbs UV, so a visible-block filter/paper combination effectively eats all light from 300 nm to 700 nm.

Aquatic biologists face a second problem, which is that the reflectance of an object depends on whether it is under water or not. There are several reasons for this, some of which we have discussed in previous chapters, but the bottom line is that the dry and wet reflectances of an object are not necessarily the same. There is also no simple relationship between the two. You can, of course, do all your measurements under water, but this is inconvenient. In addition, Spectralon is nonwettable and thus can look shiny under water as it traps air bubbles against its surface. Its underwater reflectance has been measured (Voss and Zhang, 2006), but it is no longer a perfectly diffuse reflector and thus not ideal as a standard.

Duntley (1952) came up with a simple solution; cover the object with a thin film of water and measure its reflectance in air. You have to adjust the values you get though, because the film of water itself reduces the measured reflectance. This happens because not all the light that is reflected from the object gets back out of the water film right away. Some is internally reflected multiple times, with some absorption occurring every time it hits the object. The lower the reflectance of the object, the higher its absorption and the greater the loss will be. For the usual measurement geometry where the light source is perpendicular to the object's surface and the detector is at 45°, the correction is:

$$R_{submerged} = \frac{R_w}{0.42R_w + 0.564},$$
9.1

where R_w is the measured reflectance of the wet object in air and $R_{submerged}$ is the reflectance it would actually have under water. This equation roughly doubles low reflectances and leaves high reflectances nearly unchanged. It has been extensively tested and confirmed (Petzold and Austin, 1976), but it is important to remember that the testing was all done on painted metal objects, not fish. As they say, your mileage may vary.

Mirrored and iridescent surfaces create a host of problems and are frankly a nightmare to measure. This is because the measured reflectance depends critically on both the angle of incidence and the angle of measurement. Change either by a degree and the values can go up or down by a factor of ten or more. Even worse, changing the geometry doesn't just affect the overall radiance of the measured light, but also the spectrum. In other words, you can't even assume you have the color right. If you have spent any time looking at mirrored and iridescent surfaces, this should come as no surprise. Any tilt of the object or movement of your head and everything looks different. So take any measurements of the reflectance of these sorts of objects with a terminal dose of salt. The variation is so huge that even the average of hundreds of measurements may still be useless.

A final problem is that measuring the reflectance of small objects is difficult. Fiber optic cables only get so narrow and it is difficult to position them correctly. For example, I had a student who wanted to measure the spectral reflectance of the legs of jumping spiders. These are not only small, but they are cylindrical and somewhat shiny, so fiber position is critical. After much fussing, we decided that any data we got would not be reliable. Instead, we decided to use microphotography. Essentially, my student photographed the spider legs through a microscope, taking care to include in each image a set of miniature reflectance standards, in this case a series of gray squares with known reflectances. The series of squares allowed her to create a calibration curve for each image that converted the 0–255 gray-value for each color channel (red, green, and blue) into a reflectance. A curve has to be created for each image to account for the nonlinearity of the camera (i.e., twice the gray value doesn't mean twice the radiance), and changes in lighting, exposure, camera white balance, and so forth. The obvious downside of this technique is that you only get reflectances averaged over three large wavelength bands, but it is better than nothing. If you are industrious, you can fit a monochrome camera with a rotating set of filters to cover any set of wavelength intervals you like. This creates tremendous amounts of data that can be difficult to interpret, but has been done with success (e.g., Chiao et al. 2000)

TRANSMISSION

Probably at least 90% of the measurements of biological materials are of reflectance. Many tissues are opaque, and how an object looks under reflected light is usually more relevant than how it looks under transmitted light. There are exceptions, though. The tissues of many oceanic animals transmit more light than they reflect and are often viewed from below, where transmitted light is more important than reflected light. Also, botanists, ecologists, and others care about the transmission of light through the leaf canopy. Those interested in UV damage are concerned with how the relevant wavelengths pass through the environment and through tissue. Finally, many biomedical subfields are interested in how light passes through tissue. In addition to people studying light transmission through lenses, corneas, and other ocular media, there is a rapidly growing field that attempts to diagnose skin cancers and other disorders of the dermis by measuring light transmission and scattering in tissue.

The first thing to keep in mind is that there is more than one way to define transmission. Put your thumb over the end of a working flashlight and you will see that a fair bit of red light gets transmitted. However, you cannot discern any of the details of the flashlight's bulb. All you can see is a diffuse glow. So the first big distinction is between diffuse transmission and direct transmission. Diffusely transmitted light passes through an object, but is scattered so many times that any image information is destroyed. In other words, you cannot read a book through your thumb. Directly transmitted light is either not scattered at all (i.e., it goes through a vacuum), or shows no observable scattering. Either way, image information is preserved (figure 9.7). The ratio of diffuse to direct transmission depends on the material and how thick it is. Light transmitted through window glass is mostly direct, light transmitted through a thick cloud is mostly diffuse, and light transmit-

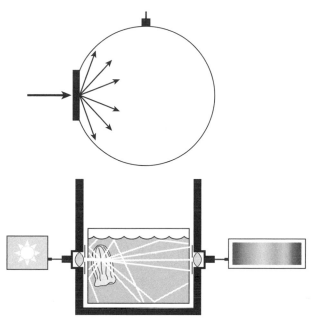

Figure 9.7: (Top) Measuring total transmission of a slablike sample using an integrating sphere. (Bottom) Measuring the direct transmission of a ctenophore in water by shining a parallel beam of light through it and then only collecting light that isn't absorbed or scattered out of the beam.

ted through a thin cloud is a mix of the two depending on the thickness. Total transmission is the sum of direct and diffuse transmission.

So, how you measure transmission depends on what you want to know. If you want to find out how bright the interior of an igloo would be, you would want to measure the total transmission of its walls. If you want to find out how visible a glass catfish would be against a background of algae, you would want to measure its direct transmission.

Whether you are measuring direct, diffuse, or total transmission, the number you get depends a lot on the geometry of the measurement system. In the ideal case, a measurement of total transmission collects all the light scattered less than 90° away from its original direction, and a measurement of direct transmission collects only light that is still going directly forward. Both of these are impossible to do. Even the best diffuse transmission setup will miss some of the light that is scattered at high angles, and even the smallest detector will still collect some light that isn't going exactly in the forward direction. There are also no guidelines. For example, there is no officially recognized cutoff angle for measuring direct transmission. Some people use detectors that collect light that has been scattered 1° or less, some 2° or less, and so on. Each geometry gives you a significantly different answer. The bottom line is that, in the real world, there is no universally accepted value for the direct transmission of an object. The same is true for diffuse transmission, but it is not as critical because the fraction of light scattered near 90° is small, so it doesn't matter much if the actual cutoff is 88° or 89°. In direct transmission, though, it matters. I usually use 1° as my cutoff (i.e., light scattered 1° or less is collected by the detector). This is totally arbitrary, though, and ultimately based on the fact that the Babylonians were big fans of the number 360 because it had so many divisors and was close to the number of days in a year.

Diffuse transmission is best measured by placing an integrating sphere directly behind the object. Then any light scattered in a roughly forward direction will enter the sphere. The original entry angle of the transmitted light gets scrambled by multiple reflections inside the sphere, so a fiber in the sphere wall measures a value that is proportional to the total amount of light that got into the sphere. This value is divided by the value you get when the object is not in the path of the beam to determine the diffuse transmission. This works wonderfully for rocks and plastics, but can get messy when you are slapping a jellyfish against the opening of your $1,500 sphere. Unfortunately, you cannot put a window on an integrating sphere, because specular reflection from the glass can change the measurement. Most integrating spheres don't come apart, either, so anything that falls into the hole might stay there for good.

Direct transmission is measured by sending a narrow parallel beam of light through your object. A detector beyond the object then collects the portion of the beam that was not absorbed or scattered more than a certain angle away from parallel. The cutoff angle depends on the diameter of the detector and its distance from the object. For small cutoff angles:

$$\theta_{cutoff} = \frac{180d}{2\pi l}, \qquad\qquad 9.2$$

where d is the diameter of the detector, l is the distance between the detector and object, and θ_{cutoff} is the cutoff angle in degrees. As with the measurement of diffuse reflection, the amount of light hitting the detector when the object isn't in the beam path serves as the reference. If you are measuring the transmission of an object that is usually under water, you must do the measurement and the reference measurement in water. Because the refractive index of tissue is much closer to that of water than of air, surface reflections are different, and overall transmission measurements are not comparable. You don't have to put the detector or light source under water, just the animal. I made a small tank out of UV-transparent acrylic that works well for this. It has a soft silicon base that you can stick pins into to trap the animal or tissue against one of the tank sides. Don't be temped to work vertically. Yes, you can just lay the tissue on the floor of the tank and shine the light vertically through it, but water surfaces wobble with the least provocation and send the light beam all over the place.

CALIBRATION

Because reflectance, transmission, and scattering are divided by a reference spectrum, you don't need to know the spectral sensitivity of the detector itself. However, if you want to measure irradiance or radiance, you must calibrate the spectrometer. For reasons that will be explained in a moment, most spectrometers do not come from the store already calibrated. Any raw spectrum from an uncalibrated spectrum not only gives no information about brightness, but also tells you nothing about spectral shape. This last part is especially critical and often overlooked. I have read many articles that give an irradiance spectrum from an uncalibrated spectrometer, justifying this by saying that they didn't need to know the absolute irradiance. This is fine, but the shape of the spectrum will also bear no resemblance to reality. This is because the sensitivity of the detectors used in spectrometers depends on wavelength (figure 9.8). The spectrometers I use are far more sensitive to green and yellow light than they are to blue and red light. It is not that the spectrometer companies use cheap stuff. All detectors have this problem, which must be fixed by calibration.

Calibration can be either relative or absolute. In relative calibration, you correct for the spectral sensitivity of the detector, but still don't know how much light you have in absolute terms. In other words, your spectrum has the correct shape, but you don't know what numbers to attach to each

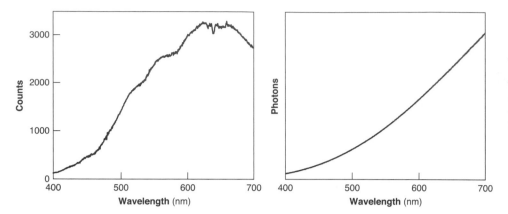

Figure 9.8: Uncalibrated spectrum of a tungsten light. Spectrum of the same light calibrated in photons. Only a relative calibration was performed, so there are no units on the y-axis. The shapes of the two spectra are quite different, due to the varying sensitivity of the spectrometer's detector to different wavelengths of light.

value—the y-axis has no labels. Relative calibration might seem like a poor cousin to absolute calibration; after all, who doesn't like to have labels on their y-axis? However, in many cases, you only care about the shape of a spectrum. Relative calibration is also easy to do. You just point a light source with a known spectral shape (usually an incandescent bulb with a known color temperature) at the detector and take a spectrum. The software packages of most spectrometers will then calculate what is called "relative irradiance." You can also create the conversion factor by hand by dividing the known spectrum of the light source by the measured raw spectrum. This "quotient spectrum" can then be multiplied by any other raw spectrum to give you relative irradiance. The nice thing about this sort of calibration is that you don't need to worry about how far the light source is from the detector, its relative angle, and so forth. As long the source light is the only light that reaches the detector, the calibration works.

Absolute calibration is more tedious. This is because you need to hit the detector with light that has a known absolute value at every wavelength. So, first of all, you need to buy a calibration source. This is a light source whose intensity is known. How is it known? Well, its intensity was measured using another calibrated spectrometer. How was that spectrometer calibrated? Using another light source whose spectrum is known. And so on. In the

United States, this chain of lightbulbs and spectrometers ends at the National Institute of Standards and Technology in Colorado (NIST), which houses the über-bulb that ultimately calibrates everything. The last time I checked, this light source was a blackbody cavity radiator that operated at a specific temperature known as the "gold point." How do they set the temperature? Well, you need to calibrate your thermometer....

The bottom line is that you need to buy an NIST-traceable light source. These cost a couple thousand dollars. Unfortunately the light source's output changes over time, so you need to send it back periodically to be recalibrated, which usually costs between $500 and $1,000. You also need to recalibrate it if you drop it, bang or otherwise fail to treat it like the rare flower it is.

In addition to having a properly calibrated source, it needs to be at the right distance from the detector. This means screwing and clamping everything down and breaking out a ruler. You may think I am kidding about screwing things down until you try this for yourself and notice how often you bump into either the detector or the source. This whole process is especially fun on a rolling ship. Fiber optic spectrometers are usually calibrated by screwing the end of the spectrometer fiber into the light source. This makes things easier, though you do have to worry about how hard you tighten the screw fitting. An extra half turn can change the numbers.

Once you have paid for and hog-tied your calibration source, the actual calibration process is simple. You shine this light of known intensity into the detector. While the details of the process depend on the spectrometer's software, and often seem unnecessarily complex, the process is relatively quick. The spectrometer creates a correction factor that converts its raw data into absolute units, generally Watts/cm^2/nm.

Even if you do everything with great care, your calibration will not be that accurate. Lightbulb drift, spectrometer temperature, and a host of other confounding variables conspire to limit your accuracy to no better than 10%. Physicists working under highly controlled conditions with more expensive equipment can get within to about 5% of the true value. This level of precision sounds lousy, and I suppose it is. After all, it is like saying that a ruler is between 11 and 13 inches long. However, few calibrated optical measurements, and none that I know of in biology, require anything close to this level of accuracy. The variation in light levels in the natural world is gigantic. One small cloud occluding the sun can drop the irradiance by a factor of ten. So a 10% error is not worth obsessing about.

Hopefully, I have convinced you that calibration is a necessary, but tedious and expensive undertaking. This may leave you wondering why spectrometers don't arrive already calibrated. You can order a calibrated spectrometer from some companies, but it is often wiser not to. This is because spectrometers are seldom used all by themselves. They nearly always have a fiber optic cable or some other optical device attached. These absorb and scatter light, which affects the calibration. In other words, you need to calibrate not just the spectrometer, but the whole measurement system. Unless you want to use your spectrometer for the same thing every time, a calibration from the factory isn't going to be of use.

One last thing: while the software in most spectrometers makes it far more convenient to calibrate before you take measurements, it is possible to calibrate afterward and use the calibration curve on your previously collected and uncalibrated spectra. You will have to do this manually in an Excel spreadsheet or similar program (Matlab, Mathematica, etc.). You will also have to keep track of the integration times you used for each uncalibrated measurement. It is best to try out this process and make sure it works before using it on data you care about.

IRRADIANCE

Once you have absolutely calibrated your system, irradiance measurements are straightforward. Remember from chapter 2 that two kinds of irradiance are the most useful, vector and scalar. Vector irradiance measures light hitting a surface from all angles in one hemisphere, but weights the amount of the light from each direction by the cosine of its angle from the perpendicular to the surface. Scalar irradiance measures all the light intersecting a surface from the whole sphere, giving all directions equal weight. Both properties are measured using special sampling optics (figure 9.9).

Most people measure vector irradiance using what is called a "cosine corrector." This is essentially a glorified white diffusing disk that scrambles the direction of the incoming light. The plastic and its exact shape have been tinkered with so that it gives the correct cosine response. Unfortunately, a cosine corrector designed for air does not give a cosine response under water (also it will leak). So, if you work in both air and water, you must buy both kinds. Cosine correctors are portable, rugged, and relatively cheap (~$200–

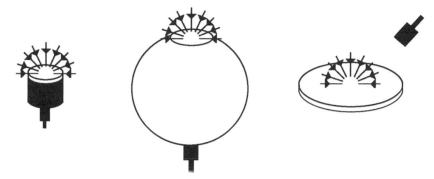

Figure 9.9: Different sampling optics for measuring irradiance. (Left) Cosine corrector, which scrambles incoming light via a small diffusing disk. (Middle) Integrating sphere, which scrambles light after it enters the top opening. (Right) Reflection from a Lambertian material (like Spectralon) that reflects equally in all directions.

$300), but have a couple of flaws. First, the cosine response is not always that good, especially at large angles from perpendicular. This usually doesn't matter much because, for many applications, the cosine corrector is facing the brightest part of the light field anyway. The bigger problem is that cosine correctors eat light. The payment for scrambling the light is that much of it never makes it through the corrector, less than 10% in many cases. This can be a problem when you are measuring dim light.

As I mentioned in chapter 2, a hole has a cosine response. So the ideal vector irradiance detector would be a hole into a chamber that scrambles the light direction. An integrating sphere is exactly this, so these have been used for a long time to measure vector irradiance. Their cosine response is close to perfect, but they still are inefficient because only a fraction of the light reaches the fiber optic detector embedded in the wall. They are also larger, more expensive, and easier to contaminate than cosine correctors. Also, because you cannot cover the hole with a window, they are difficult to use under water. Thus, while I find them useful for measurements of reflectance and diffuse transmission, I don't often use them for measuring irradiance.

A final way to measure vector irradiance is to measure the radiance of the light reflected from a perfectly diffusing reflector. As I mentioned before, a nice thing about perfectly diffusing reflectors (known as "Lambertian") is that their radiance is independent of viewing angle and linearly proportional to the irradiance striking them. In other words, look at a piece of Spectralon from any angle, and it always has the same brightness. In fact, its radiance is simply equal to the irradiance striking it divided by π. Because highly diffusing surfaces tend to reflect more light than they transmit, this system is more efficient than either a cosine corrector or an integrating sphere. It is a bit more awkward, though, because the detector is sitting above the piece of Spectralon and possibly casting a shadow. We generally get around this by using a small fiber for the detector and mounting it at a 45° angle. This method has allowed us to make irradiance measurements of skylight at much lower light levels than was previously possible.

Measuring scalar irradiance is trickier because you want to collect light from all directions, which leaves you no place to attach anything. Ideally, you would like a ball floating in space that transmits the spectral data via wireless. I haven't seen one of those yet. Instead, the usual design is a white diffusing ball at the end of a long black stick. You try to keep the stick in the darker side of the light field so it has as little an impact on the measurement as possible. The need to separate the detector from the rest of the system in this way makes the system delicate and clunky. Therefore, despite the advantages of scalar irradiance over vector irradiance for many biological applications, most people stick with vector irradiance.

RADIANCE

As with irradiance, once you have a calibrated spectrometer, radiance measurements are in theory quite simple. Restrict the view of the detector and point it at what you want to measure. The amount of measured light divided by the solid angle of the detector's field of view is the radiance.

The field of view is limited by what is called a "Gershun tube" (figure 9.10). Assuming that the end of the fiber is pretty small, the solid angle of the field of view is just the cross-sectional area of the tube divided by the square of its length. Nothing mysterious here—this is just the definition of a solid angle from chapter 2. This is in steradians. If you want to convert to square degrees, just multiply by 3283.

The Gershun tube can be a simple cardboard tube, painted black on the inside to limit reflections. However, it won't keep its shape or last long. Metal or plastic tubes painted black are better, but you do have to worry about in-

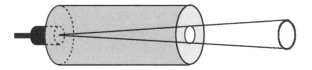

Figure 9.10: Gershun tube for measuring radiance. The field of view is restricted by a small opening in the right end of the tube.

ternal reflections. Any internal shininess, and light from outside the field of view can bounce back and forth and reach the detector. You would be amazed at how much light is reflected from black objects, especially from those that are metallic or plastic. So, if you plan on taking a lot of radiance measurements, it is best to invest in a commercial metal Gershun tube set. These have matte black interiors, internal baffles to further limit reflection, and usually come with a set of apertures that give you fields of view with standard sizes (1°, 2°, etc.). They cost about $500.

You cannot just use the bare fiber without a tube. First of all, most fiber optic cables have a huge field of view. For example, mine have a field of view that is 46° across, which translates to a solid angle of about two-thirds of a steradian or over 8,000 times the angular area of the full moon. Second, due to the laws of specular reflection, the end of the cable more efficiently collects light that strikes it head-on than at larger angles. The relationship between collection efficiency and angle also depends on whether you are under water or not. Put all this together and you are getting light from a large region with an efficiency that depends on angle, which makes measuring radiance impossible.

SCATTERING COEFFICIENT

Scattering is last because it can be tricky to measure. To do it well requires patience, experience, and rock-solid equipment that is often not cheap. Even if you do it right, your answer may not mean much because the radiance of the scattered light usually depends critically on the placement of the incident light, the object, and the detector. Turn your tissue by a fraction of a degree and things can change by a factor of ten. Avoid measuring scattering if you can.

If you are still up for it (or have no other choice), the main parameter that biologists measure is the scattering coefficient. As you remember from chapter 5, the scattering coefficient describes how much light is removed from a parallel beam as it passes through a nonabsorbing medium. The medium can be tissue, ocean water, fog, and so on, but it is usually considered an extended object. In other words, we are measuring how light is scattered within the medium, not what happens when you enter and leave it.

Measuring the scattering coefficient can sometimes be simple, but there are a host of things to consider. Let's start with the simplest case, in which the medium scatters light only moderately and absorbs hardly any. A cuvette of water with a drop of milk in it is a classic example. Send a parallel beam of light through the cuvette, measure its radiance, and call it L. Remember to do it at only one wavelength or using a spectrometer. Also remember that even the smallest detector is still finite and so will collect some scattered light as well as the direct beam. With these two variables in mind, refill the cuvette with clean water, measure the radiance again, and call it L_o. From the first equation in chapter 5:

$$L = L_o e^{-bd},$$
9.3

where d is the distance the light travels through the cuvette and b is the scattering coefficient. All this equation means is that light is being attenuated exponentially by scattering. In photon language, a fraction of the photons are being knocked out of the path by particles and thus do not make it to the detector. Solving for the scattering coefficient b gives:

$$b = \frac{\ln(L_o) - \ln(L)}{d}.$$
9.4

This is simple enough and works well for calculating the scattering coefficient of moderately turbid water, light fog, mild cataracts, and a host of other media.

There are some caveats, though. First of all, the previous two equations do not work if the scattering is so high that you get lots of multiple scattering. The equations assume that a photon, once scattered, is gone forever. However, if scattering is high enough, some of the light scattered out of the beam will get scattered back in and increase the radiance measurement, making you think the scattering coefficient is lower than it is. How high is too high? This is not easy to answer, because it depends on the angular distribution of the scattering, which depends on the medium. As a *really* rough rule of thumb, I would start to worry if my measured radiance was less than one-third of the reference radiance. Thus two-thirds of the photons are scattered at least once, which means that a fair number will be scattered more than once.

The easiest way to reduce multiple scattering is to just shorten the path length of the light through the medium. This will get you back into a regimen where single scattering dominates. However, if you do this, it is important to remember that you cannot just scale back up afterward and expect the above equations to work. In other words, you can calculate the single-scattering coefficient for skin if you choose a thin enough sample, but don't expect to plug that value into the above equations to predict how much light makes it through your torso. It won't work. It won't even be close.

I would also avoid measuring the scattering coefficient of highly transparent substances like glass and oceanic water. Unless you have an ultralong cuvette (i.e., meters long), the difference between your measured radiance and that of the reference beam will be within your typical margin of error. Also, any contamination of the medium or scratch on your containers will dramatically affect your results. There are people who devote their whole lives to perfecting the measurement of scattering in clear media, and they end up spending a large portion of their time scrubbing their glassware. Leave this tedious and thankless task to them. They are generally nice folks and happy to share their data.

So, the basic rule of thumb is to measure only the scattering coefficient if your measurement radiance is between 30% and 90% of the reference radiance. Lower than 30% and multiple scattering is a problem, above 90% and you are living in error land. These are generous limits; I usually stay between 30% and 80%.

The only remaining worry in measuring the scattering coefficient is absorption. The method described above actually measures the extinction coefficient, which is the sum of the absorption and scattering coefficients. It is equal to the scattering coefficient only when the medium does not absorb light. Unfortunately, most biological media absorb light. Even more unfortunately, a simple measurement of the radiance of your test beam will not tell you how much light is lost to each process. This is a good place to stop and think about whether you actually must know the scattering coefficient. In many cases, the extinction coefficient (which is calculated using the same equations) is good enough. Unless you are in a multiple scattering situation where scattered photons can come back to haunt you, it often doesn't matter if the photon was lost to absorption or scattering.

If you must know the relative contributions of scattering and absorption in your medium or tissue, you have to take two test measurements. The extinction coefficient measurement is just like before, but the absorption coefficient measurement is done with the beam path enclosed in a tube with mirrored interior walls. This reflecting tube ensures that nearly all the scattered light makes it to the detector by continually reflecting it back into the path. It is not perfect, because not all the scattered light makes it to the detector, and the continual bouncing of the beam increases the total path length, which increases the chance for absorption. Also, you are left with the annoying problem of how to get a reflecting tube around your sample.

I have never heard of anyone building one of these for themselves. In fact, the only example I know of is the WetLabs ac-9 transmissometer. This device, designed to measure the absorption and scattering of water at nine wavelengths, pumps the water into two tubes. A parallel beam of light is sent through each tube to a detector. One tube is black, absorbs all scattered light and thus measures the extinction coefficient c. The other is highly reflective on the inside and thus measures the absorption coefficient a. The scattering coefficient is not measured directly, but is simply the difference between c and a. The ac-9 works well, but is limited to liquids. It also costs \$40,000.

Volume Scattering Function

Unlike absorption, scattering is not completely described by its coefficient, since it also has an angular distribution. As we discussed in chapters 5 and 6, different objects and media scatter light in different directions. This is biologically quite important. For example, jellyfish and ctenophores are primar-

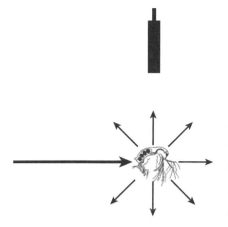

Figure 9.11: Measurement of scattering from the water flea *Daphnia* sp. A narrow parallel beam of light comes in from the left and is scattered by the phytoplankter in all directions. A Gershun tube at 90° collects light scattered in that direction and sends it to a spectrometer via a fiber optic cable.

ily visible under water because they scatter down-welling light to the side. Knowing how much light is scattered to the side would tell us how visible these animals are under different conditions.

A complete description of the scattering of an object resides in its volume scattering function, which tells us the radiance of scattered light in all directions normalized by the radiance of the incident beam. Like the bidirectional reflectance function, this is a true nightmare to measure, and I have never met anyone who has done it on a biological object more complex than the simplest of phytoplankters. However, it has been done on multicellular animals for a select number of scattering angles, most often forward, backward, and directly to the side (i.e., 0°, 180°, and 90°). These sorts of measurements have been used to optically sort cells, diagnose cancers, and predict the visibility of zooplankton (Gagnon et al., 2008).

In principle, scattering measurements are simply radiance measurements at set angles, usually normalized by the radiance of the incident beam (figure 9.11). You tightly secure your tissue or animal (in water if it is aquatic) and illuminate it with a parallel beam of light. Then you aim your Gershun tube at the object from different angles. The radiance probe should have a narrow field of view, so that it sees only the object. It also should be held quite securely. You could do all this with an irradiance cosine corrector instead of a radiance probe, but then you would have to worry about the distance between the detector and the object. Remember that radiance of an object is constant, unless there is something between it and the detector. Irradiance is not constant, and can depend on distance in non-intuitive ways.

Take these measurements, and your spectrometer will happily spit out numbers. However, you need to interpret them carefully. Before you do anything else, you might want to see what happens if you rotate your object or move your detector by a degree. In many cases, the measurement will change dramatically, possibly by an order or two of magnitude. Even repeated measurements in the same location may vary more than you would like. Angle-dependent scattering is perhaps the fussiest optical measurement you can make, particularly if you are measuring the scattering from an entire object in which surface reflections and internal refraction come into play. None of this is an artifact; it is just the way scattering of light from large objects works. This is much like the problem we had with measuring the reflectance of iridescent materials. You can, of course, average measurements over a large number of angles, but there is no reason to assume that the average is more biologically meaningful than the raw data. A shiny herring turning near the surface of the ocean will create a brief and intense reflection at some specific angle. Average out this spike and you may overlook the factor that makes it visible to predators.

So, in short, you can measure scattering at different angles, but you may have a tough time figuring out what to do with the answer.

FURTHER READING

If there were good books on measuring light, I probably never would have written my own. Everyone I know has learned how to measure light from a colleague and by trial and error. As a friend of mine from graduate school said, "Why do you think they call it RE-search?" That said, the *Handbook of Optics* (edited by Michael Bass) does have many comprehensive chapters on measuring light. Frankly, I find them heavy on equations and light on practical advice, but they do cover everything. The manuals and guides written by various optics and spectrometer suppliers can also be of use. In the end though, you are mostly left figuring it out for yourself.

CHAPTER TEN

What Is Light, Really?

"Life is weird."
"As opposed to what?"
—Found on the bathroom stall door of a truck stop in
   Breezewood, Pennsylvania

Nothing in this chapter is relevant to biologists considering light in their work, but it is interesting. Also, you might want to know what a weird substance you are dealing with.

The modern theory of light falls within the field of quantum mechanics. At first glance, quantum mechanics does not seem that strange—its name is based on the fact that light comes in units (i.e., photons) and that electrons have discrete energy states. So far, no big deal. It also includes the uncertainty principle, which states that you cannot know certain pairs of physical properties with perfect precision. Using the most often mentioned pair, the position and momentum of a photon or elementary particle, the product of the errors in measurement of both has to be greater than a certain, small number called "Planck's constant." While a bit odd, this limited precision doesn't violate most people's intuition because Planck's constant is a truly small number, about 10^{-35} J·s. Also, keep in mind that, in order to detect an object, you have to bounce something off of it. Not a problem if you are bouncing a few trillion photons off your sandwich, but bouncing a photon off an electron will affect it. Imagine trying to find a roach in a dark kitchen by swinging a baseball bat around. Once you connect with the roach, you know where it is, but you have dramatically affected its motion. There is more to it than this, but essentially the uncertainty principle comes down to the reasonable argument that measuring the position of a tiny object is going to affect it.

Quantum mechanics also involves the wave-particle duality of photons, which we have been dealing with throughout this book. This does seem odd—baseballs for example, don't seem to act like waves. But accepting Feynman's concept that photons are particles that interact in a wavelike fashion—while outside our normal experience—does not really violate our sense of logic. Well, maybe not at first . . .

This chapter explores two of the most unusual aspects of quantum mechanics, two-slit interference and quantum entanglement. Both violate our most fundamental notions about how the world works. So far as I know, no animals take advantage of either of these aspects, but these are things that you really need to see before you die, just like the Grand Canyon. As with the canyon, the view is marvelous and uplifting, but only because it's a mighty big hole.

THE TWO-SLIT EXPERIMENT

If you are a practicing biologist, you probably took two semesters of college physics, the second of which was called "Electricity and Magnetism" and had a few optics labs. One of the labs likely included a demonstration of what is called "two-slit interference." The setup would have been simple—a light source (usually a red laser) aimed at a piece of cardboard bearing two narrow rectangular slits (figure 10.1). Behind the two slits, there would be another piece of cardboard (or just the wall), upon which you would see a pattern of red and black stripes. If you had an especially nice lab, you would be able to increase the separation of the slits and notice that this decreased the distance between the stripes. You could write up a report saying that the stripes were the interference pattern of two spherical wavefronts exiting the slits, mention how this showed the wave nature of light, and go home.

But what if you did the experiment one photon at a time? This is not as hard as you might think. As I mentioned in chapter 9, photomultiplier tubes (PMTs) are capable of detecting single photon strikes. In fact, Einstein's discovery of the photoelectric effect underlying PMTs was the first proof of the quantal nature of light and the stated reason for his Nobel Prize. Photon-counting PMTs cost a bit extra, but they are common lab equipment for physicists. Extremely dim light sources that emit one photon at a time are less common, but far from exotic. So, if you replace the laser with a one-photon-at-a-time light source and the wall with a bank of photon-counting PMTs, what do you see? Your intuition might tell you that you would see

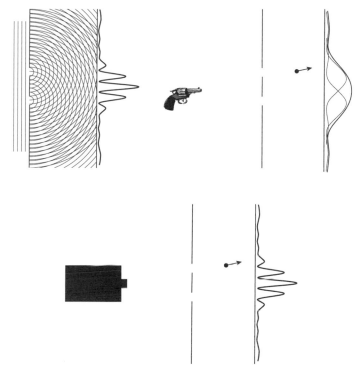

Figure 10.1: (Upper left) Classical description of two-slit interference, where a beam of light goes through two small slits. The spherical expanding wavefronts on the other side interfere with one another and create a pattern of light and dark stripes. (Upper right) A madman with a pistol shooting wildly at a wall with two slits will leave two collections of bullets in the wall beyond the slits. The total probability of a bullet hitting any part of the wall forms a smooth curve with a single broad peak. (Bottom) Performing the two-slit experiment with a light source that only emits one photon at a time. Rather than forming a smooth curve as was seen with the pistol, we instead see an interference pattern.

two piles of detected events, centered behind each slit, much like you would see if someone blindly shot bullets at a wall with two holes.

Instead, you see the same diffraction pattern as before, even if the photons are emitted days apart. It takes awhile for the pattern to appear, because the photons are arriving so slowly, but after a while it will look just like the classic interference pattern you saw in college, suggesting the photon somehow interfered with itself.

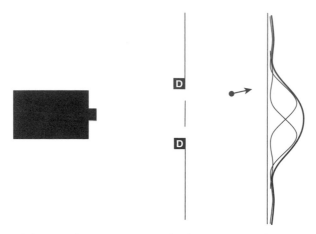

Figure 10.2: The two-slit experiment with photons emitted one at a time, but now with detectors that record which slit the photon went through. The interference pattern is no longer seen.

Defining the size of a photon is a tricky thing, but you would probably wager that it is smaller than the distance between the two slits. The fact that photons are absorbed by single photoreceptors in our eye should give you an idea of how small they are. So, based on our human sense of how the world should work, the photon must have gone through either one slit or the other. Is there a way to find out which?

It turns out there are instruments that can detect a photon without completely absorbing it like a PMT does. If you put one of these instruments in front of each slit, one or the other (at random) will signal that it has detected a photon. In other words, the detectors will tell you that each photon has gone through one slit or the other. But, if you take this measurement, the interference pattern behind the two slits will vanish, being replaced by a pile of photon detections behind each slit, just as if you were throwing baseballs through two holes at the state fair (figure 10.2).

So ... it looks like a single photon, however small, goes through both slits and interferes with itself to create the striped pattern you see. However, if you measure which slit it goes through, you will get an answer but the interference pattern will go away. In other words, peek inside the magic box and it breaks.

If this doesn't bother you on a fundamental level, call me. We have explored this world for thousands of years and discovered many puzzling things, but this is the first that truly violated our sense of logic. Even the odd results of special relativity were logically derived and make their own sort of sense.

Most philosophers begin with the principle that the laws of logic are external to ourselves and self-evident, but here we have a clear case of a solid experimental result that has no logical explanation. As a biologist, I might suggest that our rules of logic are not universal, but simply the result of natural selection for the mental ability to predict the motion and interactions of large objects. Maybe there is a species out there that is concerned with the motions of photons and for whom this experiment makes intuitive sense. Or maybe it's weird for everyone.

So how do the physicists deal with it? On an intuitive level, most whom I have met subscribe to some variant of the Copenhagen interpretation, developed in 1927 by two of the pioneers in the field—Niels Bohr and Werner Heisenberg (who was visiting Bohr in Copenhagen). There are long books on the history and philosophy of this set of principles, filled with controversy, but at their core, they say the same thing, "What a photon does between emission and detection is its own business." While the emission of photons and the detection of photons are observable events, assigning meaning or metaphor to what a photon is doing in transit is pointless. The equations developed for waves are fairly good at predicting what you will measure in a set of detectors (e.g., an interference pattern), but light in transit is not some evanescent water wave. Neither is it anything else. The math will give you the right answer for any experiment you can devise (and so far, it has, to amazing precision), but there is no meaning to a photon in transit.

It is important to realize that this interpretation is not saying we don't know what a photon is doing in transit, but rather that it cannot be known. If you try to find out (for example, using the detectors in the two-slit experiment), you will wreck things. On one hand, this is seriously unsatisfying. However, with time it develops a sense of purity in the "If you love photons, set them free" sort of way.

To avoid the wave metaphor and to deal with the fact that a photon appears to interfere with itself, quantum physicists have adopted the concept of the probability amplitude. A probability amplitude is a probability, except that it also has phase like a wave does. To determine what happens in any particular experiment, one first adds up the probability amplitudes for all possible paths, remembering to account for the phase using exactly the same math we would use if we thought light was a wave. Then you square the amplitude of the sum to get the final probability of the photon appearing in a certain place (just like we square the height of a water wave to determine the energy in it). If you did this for the two-slit experiment (remembering that the math is just like wave math), you would find that the most probable places for a photon to land on the wall are right where we see the interference stripes.

You are probably thinking, "This is lame, all you've done is substitute 'probability amplitude' for 'wave.'" However, "probability" amplitude is a neutral word without the metaphorical baggage of waves. It also allows you to think of situations in which there is only one photon at a time. More importantly, though, it allows you to include things that waves never do. For example, light doesn't always travel at the speed of light; it has a probability amplitude to travel faster and slower. It also has a probability amplitude to travel backward in time and to turn into a positron and electron (matter and anti-matter) that will annihilate each other shortly thereafter and re-form the photon. In short, a photon going from the emitter to the wall behind the two slits has a probability amplitude to be anywhere in the universe, anywhere in time, and the potential to form as many electron-positron pairs as it likes. The squared sum of all these possibilities results in the familiar interference pattern you see on the wall.

You might think this is all metaphysics, but to get the correct final probability, you must take every one of these infinite possibilities into account. Luckily, the probability amplitudes decrease as the paths get more baroque, but it is a still a nightmarish calculation, even for the simplest case of one electron interacting with one photon. This particular case has been worked out in great detail, because it can be compared with a physical property than can be measured with huge precision, the electron g-factor (essentially the interaction between an electron and a magnetic field). Classical theory says that the g-factor should be 1, but experiments show that it is actually 1.0011596521808576, with some uncertainty in the last two digits (Odum et al., 2006). The most recent theoretical value I know of (which includes up to eight complicating events), is 1.00115965246 (Feynman, 1985). The two differ by less than one part in a billion. Pretty good agreement, especially compared to what biologists usually see in their own data.

There are, however, other interpretations of the two-slit experiment. One is that the photon goes through both slits, but in different universes. In other words, every time a measurement is made, the universe splits in two, one universe for each answer (Hugh Everett III, 1957).This is known as the "many worlds" interpretation, though "many" is a serious understatement. As the physicist Paul Davies said, this theory is "cheap on assumptions, but expensive on universes." While some research has suggested that this theory is logically consistent (e.g., Deutsch, 1998), most physicists I have met do not feel it is a useful idea.

A second interpretation is that photons are not a useful or accurate way to describe light. After all, nearly every concept in this book was presented by considering light to be a wave. Subjects like scattering and interference can be explained using the photon concept, but it is messier and far less intuitive. The only place in which the quantal nature of light becomes obvious is in its interaction with matter. As we mentioned above, a sensitive light meter will record individual events, leading us to think of photons as insects flying into a bug zapper. However, just because something interacts with matter in a quantal fashion doesn't mean that the thing itself is packaged in discrete units. Suppose you shake a crib with a sleeping baby. If you shake it hard, the baby always wakes up. However, if you shake it gently, the baby might wake up. The waking up itself is a quantal event—the baby is either awake or asleep—but the probability of this happening depends on how hard you rock. As a child, I remember a similar process working at the dinner table with my little brother. Kick him lightly under the table long enough and he would eventually blow his stack. You never knew quite when, though.

Moving back to the two-slit experiment, perhaps the dim light that leads to single and separated events at the detector is just a weak wave with a low probability of triggering a detection event. Then the behavior of the experiment makes sense. Everyone expects a wave, even a weak one, to interfere with itself after passing through two slits. The only sticking point is the assumption that the light is traveling in individual packets.

I find this idea (reviewed by Kidd et al., 1989) appealing, since it dispenses with much of the mystery of quantum mechanics. It does leave the question of why the interaction of light waves with matter is quantal, but—to my mind at least—this is less disturbing than objects being in two places at once or creating uncounted numbers of universes before breakfast. Personally, I have always liked the idea of photons, imagining them as little colored balls that I could keep on my nightstand, if only they would stay still. However, since both photons and light waves are concepts (we will never see either), I would prefer to use a concept that works most of the time and doesn't violate my intuition. This is just my preference, though, and I am dancing lightly over a complex and subtle field. To learn more about the sorts of two-slit experiments that have been done and how they have been interpreted, start with Feynman's *QED* and work your way through the vast popular and scientific literature on the subject.

QUANTUM ENTANGLEMENT

Possibly the strangest aspect of quantum mechanics is what has been termed "entanglement." While the two-slit experiment can be at least partially explained via various intuitive arguments, the behavior of entangled photons leaves everyone perplexed. As David Mermin once said, it is about as close to magic as one can get in physics (Mermin, 1985). Mermin also quotes an unnamed physicist as saying that anyone who isn't bothered by it has to have rocks in their head (Mermin, 1985). Entanglement is a fairly general phenomenon that also exists in elementary particles. However, since this is a book about light, I'll stick to entangled photons. Just keep in mind that what follows is not specific to light and polarizers, but a general aspect of the effects of measurement. In other words, it is central to everything.

So, to begin at the beginning, certain photochemical processes produce pairs of photons with correlated polarizations. In other words, the difference between the angles of polarization of the two photons is constant. The exact difference depends on how the photons are produced, but is not critical to the central concept of entanglement. For example, a photon that enters a crystal of beta barium borate (β-BaB_2O_4) will sometimes be split into two photons, each with half the energy and double the wavelength of the original photon (Dehlinger and Mitchell, 2002). The process, known as "spontaneous parametric down conversion," does not appear to be relevant to biology, so we'll just treat it as a black box. These new photons leave the crystal traveling in opposite directions but have the same polarization. They are said to be entangled.

However, while you know that the polarizations of the two photons are identical, you do not know what they are. They could both be vertically polarized, horizontally polarized, or at any angle in between. Let's see what happens when you try to find out.

Imagine you have built a contraption that shines 400 nm violet photons one by one into a beta barium borate crystal and creates some entangled pairs of 800 nm photons. For convenience, let's assume that one photon of each entangled pair goes to the right and the other one goes to the left. We'll detect them with photon-counting photomultipliers.

For now, let's just put a detector in the path of the right-traveling photons. Turn the detector on and you will detect a steady stream of them. The number will be much smaller than the number of violet photons that originally entered the crystal, because only a small fraction of them get split—the rest just pass right through. Now place a polarizer between the crystal and the detector. The number of photons you detect now drops by 50%. Turning the polarizer doesn't make any difference—you always get half the photons you had before. This makes sense, because we said that the polarization of the photons was random, and we know from chapter 8 that a randomly polarized photon passes through an ideal polarizer exactly 50% of the time, no matter how the polarizer is oriented. Move the detector and the polarizer to the other side of the crystal to measure the left-traveling photons and you will get the same result.

Bring the polarizer and detector back to the right side and turn the polarizer so that it is vertical. Just as before, half the photons get through. Now get a second detector and vertically oriented polarizer and put it in the path of the left-traveling photons. To make things simpler to describe, let's have each detector flash a green light when it detects a photon, and a red light when it doesn't (figure 10.3). Turn everything on. Now you notice that every time the right detector sees a photon, the left detector also sees one, and both lights flash green. Also, every time the right detector doesn't see a photon, the left detector also doesn't see one and both lights flash red. You never see a mix of green and red lights. You do however notice that you see double-red as often as you see double-green, within the usual statistical variation.

Seems innocuous, but think about it for a moment. Suppose the right-traveling photon always had a vertical polarization. Then the left-traveling photon always would as well and they would both get though a vertical polarizer. If that were true though, then the lights would flash double-green all the time. However, we see that half the time they flash double-red, meaning that no photon was passed by either polarizer. So they cannot both be vertically polarized.

Maybe the two photons were both polarized at 45°? From Malus's law in chapter 8, they should then pass through a vertically oriented polarizer 50% of the time ($= \cos^2 45°$). This seems promising. However, the chance of both photons making it through their respective polarizers would be

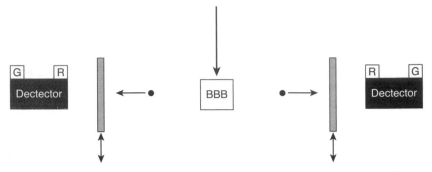

Figure 10.3: Schematic of the entanglement experiment. Violet photons (400 nm) enter the beta barium borate (BBB) crystal at center. Some of these photons are split into two 800 nm photons that have the same unknown polarization. One travels to the left and one to the right. Each intercepts a polarizer oriented so that it maximally passes photons with vertical polarization. Behind each polarizer is a detector that flashes green (G) when it detects a photon, and flashes red (R) when it doesn't.

$50\% \times 50\% = 25\%$. Similarly, the chance for neither getting through is also 25%. So the chance of the lights flashing either double-red or double-green would be 50%, not the 100% we are seeing. You can try other possibilities for the polarization angle of the photons if you like, but you'll find that none of them will give you 50% double-red, 50% double-green and no mixing or red and green lights.

So it looks like the passage of one photon through its polarizer somehow forces its entangled partner to pass through its polarizer. Similarly, the non-passage of one photon forces the nonpassage of the other. Having both polarizers at the same angle is just a specific case. In general, they could have any angle relative to each other, though this changes how the passage of one affects the other. For example, if one polarizer is vertical and the other is horizontal, then if one photon makes it through the vertical polarizer, the other photon will never make it through the horizontal one. In its most general form, the passage of one photon through a polarizer changes the odds of the entangled photon passing through its polarizer from one-half to $\cos^2\theta$, where θ is the difference in the angles of the two polarizers. Even stranger, this forcing effect appears to be instantaneous, or at least the effect travels much faster than the speed of light. Careful experiments over distances of many hundreds of meters have proven this (Salart et al., 2008). This has nothing to do with the nonlinear crystal, which was just a convenient way of producing entangled photons. It also is not specific to light, though it is easiest to demonstrate with photons and polarizers. At its heart what we are saying is that the measurement of one object can affect the measurement of another, instantaneously and possibly across vast distances. You can see why Einstein referred to this as "spooky actions at a distance."

The only way to avoid this spookiness is to assume that photons carry instruction sets that tell them what to do for any orientation of the polarizer. For example, "If the polarizer is set to 45°, always go through, if it is set to 0°, never go through," and so on. The instruction sets would in general be different for different photons (i.e., photons would come in flavors), but the two photons of an entangled pair would carry the same instructions. This is known as the "hidden variables" hypothesis. It is a strange idea, but strange and true are kissing cousins in quantum mechanics. Also, no one can think of any other idea that preserves a remotely intuitive sense of reality. See if you can come up with one.

This debate between spooky actions and hidden variables was introduced by Einstein, Boris Podolsky, and Nathan Rosen in 1934 (Einstein et al., 1934) and quickly became known as the "EPR paradox." It remained a philosophical discussion until John Bell found an experimental way to distinguish the two possibilities (Bell, 1964). The paper is only six pages long, but not terribly accessible to nonphysicists. However, the core ideas are straightforward and have been beautifully explained by others. The following explanation closely matches that given by David Mermin (1985).

We'll use our original apparatus with its violet light, nonlinear crystal, and pair of polarizers and photo-counting detectors with their red and green lights, but add a device to each of our two polarizers that randomly rotates them to one of three orientations: 0° (vertical), 120°, or −120°. The orientations of the two polarizers are independent of each other and selected randomly each time a new violet photon is produced by the light source (figure 10.4). The exact timing of the orientation changes doesn't matter, as long as it happens before the photons pass through them. To keep things simple, we will assume that photons are emitted at a steady rate and that each photon entering the crystal gets split.

Turn everything on and run a bunch of photons through the system. You will find that, on average, the lights flash red-red, green-green, red-green, and green-red in equal proportion (each one-quarter of the time). In particular, you will find that the two lights flash the same color half the time and different colors half the time. However, you will also see that, whenever the two

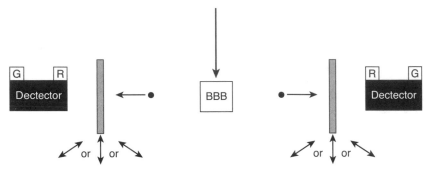

Figure 10.4: The apparatus in figure 10.3 modified such that each polarizer is randomly turned to one of three positions: 0° (vertical), 120°, or −120°.

polarizers have the same orientation (both vertical, both 120°, or both −120°), the two lights will always flash the same color. When the angles of the two polarizers are different, the lights flash the same only one-quarter of the time. Both of these conditions can be explained by the cosine-squared law described above. One-third of the time, the two polarizers are oriented at the same angle, so $\cos^2\theta = 1$, meaning that either both photons get through or neither does. The remaining two thirds of the time, the polarizers are oriented at different angles. In this case, the angle between them is always either 120° or −120°, so $\cos^2\theta = \frac{1}{4}$ and the lights flash the same color only one-quarter of the time. Add the two cases together and you get $\frac{1}{3}\cdot 1 + \frac{2}{3}\cdot\frac{1}{4} = \frac{1}{2}$, meaning that the lights flash the same color half of the time, which matches what we see. So far, so good.

Let's assume that the hidden variables hypothesis is true. There are three polarizer orientations, so an instruction set would have to tell the photon what to do for each. For each orientation, a photon could either pass through or not, which means that there are eight possible instruction sets ($= 2^3$). A possible instruction set could be to always pass through the vertical polarizer, always pass through the 120° one, and never pass through the −120° one. Let's call this instruction GGR, shorthand for the green, green, and red lights that would flash for polarizers oriented at 0°, 120°, and −120°, respectively.

Since two entangled photons always flash the same lights whenever both polarizers have the same orientation, they must have the exact same instruction sets. However, this gets us into trouble with the fact that the two lights flash the same color 50% of the time. Let's see why. Two of the instruction

TABLE 10.1

Results of instruction set GGR for all nine possible combinations of polarizer positions for the left and right polarizers. The results are in bold when the two colors are the same.

Left	Right	Colors (left/right)
0	0	**GG**
0	120	**GG**
0	-120	GR
120	0	**GG**
120	120	**GG**
120	-120	GR
-120	0	RG
-120	120	RG
-120	-120	**RR**

sets, namely RRR and GGG, flash the same color every time. What about the remaining six? Let's again take GGR for an example. There are nine possible pairs of orientations of the polarizers (table 10.1). You can see that, for this rule set, the two lights flash the same color 5/9 of the time. The underlying reason for this is that one color (green) occurs twice in the instruction set and the other occurs only once. The other five rule sets will have the same problem—one color will always appear twice and the other only once. You can try them if you want, but you will always find that the lights will flash the same color 5/9 of the time. Combine this with the fact that photons with the instruction sets RRR and GGG always flash the same color (there is only one color to flash), and you find that, if the two entangled photons have the same instruction set, then the two detectors should flash the same color at least 5/9 of the time, which is more than the 50% that is observed.

This experiment has been carried out many times in many labs (even in undergrad teaching labs) and the end result is that the same colors flash significantly less than 5/9 of the time (usually by at least ten standard deviations). Therefore, the hidden variable hypothesis cannot be true. This means that the mere process of passing one photon through a polarizer affects the polarization of a correlated but otherwise seemingly independent photon at great distances. In other words, the properties of an object depend somehow on what is going on in other places.

The experiment has been performed at increasingly longer distances. The current distance record is at least 50 kilometers, using photons sent through optical fibers (Marcikic et al., 2004). In open air, the photons get affected by things along the way, so the record is shorter, but still at least 600 meters (e.g., Aspelmeyer et al., 2003). Other experiments have tested the speed of the effect and have found that it is at least 10,000 times the speed of light (Salart et al., 2008).

If nothing else about light bothers you, quantum entanglement really should. Though, as I said before, it is not just about light, but about all matter. Similar experiments have been performed using entangled elementary particles whose spins are measured using magnets. The experiments are harder to do, but give the same results. It is also not about polarizers or magnets, but about the process of measurement itself. We started this chapter with the uncertainty principle, which says that you cannot measure the properties of small objects with arbitrary precision. I said that this was understandable because the measurement itself would disturb the object (the roach-bat analogy). But how does this explain why measuring one photon affects another photon miles away?

At the moment, there is no answer to this. One possibility is that there is some unknown force that travels from a measured photon to its entangled partner at a speed of at least 10,000 times the speed of light. The other possibility is that space doesn't exist, at least not in the way we think it does. While we may think of the two entangled photons as being miles apart, in some more fundamental way they may be in the same spot. I personally prefer this hypothesis. It's clean and perverse enough to be true. It also, to my mind at least, has a nice symmetry with the theory of special relativity. One of the central tenets of relativity is that you cannot truly say two events are simultaneous. Instead, their relative timing depends on the motion of the viewer relative to the two events. Perhaps in quantum mechanics we can't say whether two things are truly in the same place or not.

As long as we're on relativity, it's important to mention that, as perverse as entanglement is, it doesn't violate the principles of special relativity. It's true that the influence of measurement travels far faster than the speed of light and may in fact be instantaneous. But you can't use it to send a signal. Suppose you are sitting next to one of the detectors of our gadget and your friend is miles away sitting next to the other detector. If you could control whether the photon makes it through your polarizer or not, you would affect the photon that your friend sees. Learn Morse code and you would be able to chat at speeds substantially faster than the velocity of light. Unfortunately, you can't control whether the photon goes through your polarizer. On average it makes it through half the time. For any given photon, it's a flip of the coin. So you can't send a deliberate signal, and thus relativity is not violated.

However, because measurement affects entangled photons, you can use them to tell if someone has been eavesdropping in on a conversation. This is the basis for what is known as "quantum cryptography"—a bit of a misnomer, because the system is not a code, but a way of determining whether a message has been tampered with. The most secret codes can only be broken if you have the key. The problem then is safely transmitting the key. You could encrypt the key, but then you have to worry about safely transmitting the key for the key, and so on. In practice, it is impossible to send a key in a completely safe way. However, you can do the next best thing. Using a system of entangled photons, you can at least determine if someone tried to read the key during transit. Any attempt to intercept the transmission is essentially a measurement and will affect the correlations between the polarizations of entangled photons. If a change is noticed, the key is thrown out and another attempt is made to send it. As far as anyone knows it is impossible to intercept a key sent via quantum cryptographic methods without its being noticed.

These systems exist. Since about 2004, certain high-level bank transfers and small computer networks use quantum cryptography, and several companies market the systems. The current distance records are about 200 kilometers for fiber-based systems and 150 kilometers for open-air systems. I personally don't have any secrets that would require tearing the fabric of the universe to discover, but it never fails to amaze me how even the most esoteric science can end up being marketed via companies with names like ID Quantique, MagiQ Technologies, SmartQuantum, and Quintessence Labs. Obviously, it's all about the Q.

THE END OF THE ROAD

So we have come a long way from our original concerns with units and geometry. Perhaps more than any other field of science, our desire to understand light has opened up entire worlds, from Newton's discovery of the colors within white light, to the nineteenth-century research that ultimately showed that electricity, magnetism, and light were all closely related, and finally to the mind-altering twin upheavals of special relativity and quantum mechanics. There are those that, impressed by the numerical accuracy of quantum mechanics, feel that this journey, however exciting, has now come to an end.

I suppose it's possible. While new things about light are always being discovered, nothing as fundamental as quantum mechanics has appeared for many decades, despite an unprecedented number of smart and well-funded people working in the field. However, I cannot help but hope that another simple insight could open another new world. After all, Einstein's theory of relativity developed out of a simple idea he had— "If I put a mirror on my bike and ride fast enough, will my reflection eventually disappear?" Experiments are of course critical as well, but never doubt the power of a simple idea to upturn the world.

There was a book I read as a child. I don't remember the title or even what happened in it. All I know is that it had a boy who could enter the fourth dimension. He didn't need any special equipment; he simply turned in the right direction. I think of him often and imagine the vast unseen world that must be inches from our faces and minds, hidden until one of us learns to look in just the right way.

FURTHER READING

Numerous books have been written about quantum weirdness, but my favorite is still Richard Feynman's *QED*. It's short and simple, but covers the important facts. It also steers clear of a lot of the whacky metaphysics that has grown up around the field. David Mermin's 1985 paper on the EPR paradox is my favorite treatment of the subject, avoiding a lot of the complications that are seen in other papers. Read and digest these two short works and you will know as much about quantum weirdness as almost anyone.

Converting Spectral Irradiance to Lux

You can convert an irradiance spectrum $E(\lambda)$ given in W/m²/nm to lux using:

$$\text{lux} = 673 \sum_{\lambda=380}^{\lambda=780} E(\lambda)\bar{y}(\lambda)\Delta\lambda, \qquad \text{A.1}$$

where $\bar{y}(\lambda)$ is the photopic (light-adapted) luminosity curve for humans (given in the table below). In other words, to calculate lux, you multiply your irradiance spectrum (first binned into 10 nm intervals) by the luminosity curve and by $\Delta\lambda$ (which is 10 nm in this case) and then add up all the products. Then multiply this final sum by 673. This is most easily done in a spreadsheet program. Remember that your irradiance must be in watts, since that's how the luminosity curve was created. The values of $\bar{y}(\lambda)$ below 400 nm and above 700 nm are small, so don't worry if you don't have irradiance data for this range.

wavelength (nm)	$\bar{y}(\lambda)$	wavelength (nm)	$\bar{y}(\lambda)$
380	0.000039	590	0.76
390	0.00012	600	0.63
400	0.00040	610	0.50
410	0.0012	620	0.38
420	0.0040	630	0.27
430	0.012	640	0.18
440	0.023	650	0.11
450	0.038	660	0.061
460	0.060	670	0.032
470	0.091	680	0.017
480	0.14	690	0.0082

wavelength (nm)	$\bar{y}(\lambda)$	wavelength (nm)	$\bar{y}(\lambda)$
490	0.21	700	0.0041
500	0.32	710	0.0021
510	0.50	720	0.0011
520	0.71	730	0.00052
530	0.86	740	0.00025
540	0.95	750	0.00012
550	0.995	760	0.000060
560	0.995	770	0.00030
570	0.95	780	0.00015
580	0.87		

While the above procedure always works, you can sometimes get away with a simple conversion if the light source is fairly monochromatic, such as a light-emitting diode, or light that has passed through an interference filter. In those cases, the table below is useful. It takes one W/m^2 of monochromatic light and converts it to various common irradiance units as a function of wavelength. For example, 1 W/m^2 of 500 nm light equals 2.52×10^{18} photons/s/m^2 and 215 lux. The table assumes that you are given the initial value in W/m^2, because this is what spectrometers are most likely to do, and what lighting engineers and light-source spec sheets usually provide. The wavelength dependence of the conversions is much more pronounced for converting to lux and foot candles than it is for going from watts to photons. So, while you can get away with converting the average bioluminescence spectrum from watts to quanta using this table, converting to the two photometric units will give you the wrong answer. Instead, you will need to do the integral described above.

wavelength	photons/s/m^2	microEinsteins/s/m^2 ($=\mu mols/s/m^2$)	lux ($=lumens/m^2$)	foot candles ($=lumens/ft^2$)
400	2.01×10^{18}	3.35	0.269	0.0250
420	2.12×10^{18}	3.51	2.69	0.250
440	2.22×10^{18}	3.68	15.5	1.44
460	2.32×10^{18}	3.85	40.4	3.75

wavelength	photons/s/m²	microEinsteins/s/m² $(=\mu mols/s/m^2)$	lux $(= lumens/m^2)$	foot candles $(= lumens/ft^2)$
480	2.42×10^{18}	4.01	94.2	8.75
500	2.52×10^{18}	4.18	215	20.0
520	2.62×10^{18}	4.35	478	44.4
540	2.72×10^{18}	4.52	639	59.4
560	2.82×10^{18}	4.68	670	62.2
580	2.92×10^{18}	4.85	586	54.4
600	3.02×10^{18}	5.02	424	39.4
620	3.12×10^{18}	5.18	256	23.8
640	3.22×10^{18}	5.35	121	11.3
660	3.32×10^{18}	5.52	41.1	3.81
680	3.42×10^{18}	5.69	11.4	1.06
700	3.53×10^{18}	5.85	2.76	0.256

Calculating the Absorbance Spectrum of a Visual Pigment

A number of functions approximating the absorbance curves of visual pigments have been proposed over the years. This one, by Govardovskii et al. (2000), is the most commonly used and fits the experimental data well. It also does a good job of estimating the sensitivity to light at wavelengths far longer than the peak wavelength of absorption, which is good if you want to determine whether a deep-sea fish with a 470 nm pigment can see a 680 nm red light.

The visual pigment absorbance spectrum is the sum of its alpha and beta bands:

$$A(\lambda) = \alpha(\lambda) + \beta(\lambda). \tag{B.1}$$

Let $\lambda_{max,\alpha}$ be the peak wavelength of the alpha band. Then, for a retinal-based (A_1) visual pigment, the alpha band is well approximated by:

$$\alpha(\lambda) = \left(e^{69.7(a-x)} + e^{28(0.922-x)} + e^{-14.9(1.104-x)} + 0.674 \right)^{-1}, \tag{B.2}$$

where $x = \lambda / \lambda_{max,\alpha}$, and $a = 0.8795 + 0.0459 e^{-(\lambda_{max,\alpha} - 300)^2 / 11940}$.

The beta band is well approximated by:

$$\beta(\lambda) = 0.26 e^{-((\lambda - \lambda_{max,\beta})/b)^2}, \tag{B.3}$$

where $\lambda_{max,\beta} = 189 + 0.315 \lambda_{max,\alpha}$ and $b = -40.5 + 0.195 \lambda_{max,\alpha}$.

Similar equations exist for A_2-based pigments (see Govardovskii et al., 2000, for details).

Remember that these equations are just glorified exercises in curve-fitting. They do a good job of fitting any A_1-based visual pigment that has a

λ_{max} between 330 nm and 600 nm, but the forms of the equations say nothing about the actual physics. The fact that you only need to know the λ_{max} to determine the whole absorbance curve probably says something interesting about the underlying mechanisms of absorption, but these equations don't directly address that.

Refractive Indices of Common Substances

	Absolute refractive index	Reference
Organic		
human vitreous humor	1.336	Sivak and Mandelman, 1982
cytoplasm	1.35	Charney & Brackett, 1961
human cornea	1.37-1.40	Sivak and Mandelman, 198
human lens	1.38-1.40	Sivak and Mandelman, 1982
mitochondria	1.40	Beuthan et al., 1996
cell membrane	1.46 to 1.6	Quinby-Hunt and Hunt, 1988
lipids	1.48	Beuthan et al., 1996
protein	1.55	Chapman, 1976
Inorganic		
air	1	*CRC Handbook of Chemistry and Physics*
water/ice	1.33/1.31	*CRC Handbook of Chemistry and Physics*
cryolite	1.338	*CRC Handbook of Chemistry and Physics*
fused silica	1.39-1.42	Aas, 1981
calcite	1.49 and 1.66	*CRC Handbook of Chemistry and Physics*
quartz	1.55	*CRC Handbook of Chemistry and Physics*

Optical Properties of Very Clear Water

Because water is central to biology, I have included a table of its absorption coefficient (a), scattering coefficient (b), real refractive index (n), and imaginary refractive index (k). The absorption and scattering coefficients a and b were measured by Smith and Baker (1981) in very clear but naturally occurring bodies of water. The complex refractive indices were measured using very pure water in a lab (see Segelstein, 1981, for details).

wavelength (nm)	$a\ (m^{-1})$	$b\ (m^{-1})$	n	$k\ (x\ 10^9)$
300	0.1410	0.0263	1.371	4.148
310	0.1050	0.0228	1.368	3.546
320	0.0844	0.0199	1.365	3.190
330	0.0678	0.0174	1.363	2.984
340	0.0561	0.0153	1.360	2.766
350	0.0463	0.0135	1.358	2.528
360	0.0379	0.0119	1.356	2.316
370	0.0300	0.0106	1.354	2.117
380	0.0220	0.0095	1.353	1.940
390	0.0191	0.0085	1.351	1.761
400	0.0171	0.0076	1.350	1.580
410	0.0162	0.0068	1.348	1.422
420	0.0153	0.0061	1.347	1.258
430	0.0144	0.0055	1.346	1.088
440	0.0145	0.0050	1.345	0.939
450	0.0145	0.0046	1.344	0.809
460	0.0156	0.0041	1.343	0.760
470	0.0156	0.0038	1.342	0.729
480	0.0176	0.0034	1.341	0.709
490	0.0196	0.0032	1.340	0.734
500	0.0257	0.0029	1.339	0.924

wavelength (nm)	a (m⁻¹)	b (m⁻¹)	n	k (x 10⁹)
510	0.0357	0.0027	1.339	1.267
520	0.0477	0.0024	1.338	1.570
530	0.0507	0.0022	1.337	1.757
540	0.0558	0.0021	1.337	2.098
550	0.0638	0.0019	1.336	2.442
560	0.0708	0.0018	1.335	2.869
570	0.0799	0.0016	1.335	3.434
580	0.1080	0.0015	1.334	4.434
590	0.1570	0.0014	1.334	6.365
600	0.2440	0.0013	1.333	9.634
610	0.2890	0.0012	1.333	12.380
620	0.3090	0.0011	1.332	13.990
630	0.3190	0.0011	1.332	15.020
640	0.3290	0.0010	1.331	15.700
650	0.3490	0.0009	1.331	16.740
660	0.4000	0.0009	1.330	19.400
670	0.4300	0.0008	1.330	20.980
680	0.4500	0.0008	1.329	23.000
690	0.5000	0.0007	1.329	26.530
700	0.6500	0.0007	1.329	33.480
710	0.8390	0.0006	1.328	49.980
720	1.1690	0.0006	1.328	72.910
730	1.7990	0.0006	1.328	115.000
740	2.3800	0.0005	1.328	145.800
750	2.4700	0.0005	1.327	155.900
760	2.5500	0.0005	1.327	158.000
770	2.5100	0.0004	1.327	152.700
780	2.3600	0.0004	1.326	140.900
790	2.1600	0.0004	1.326	128.200
800	2.0700	0.0004	1.326	125.000

Optical Properties of Natural Waters

While the optical properties of pure water are useful, the properties of natural bodies of water are quite different due to the presence of particles and dissolved absorbing substances. The great optical oceanographer, Nils Jerlov, developed a classification scheme in the early 1950s that is still used today. The system includes five types of ocean water, where any extra absorption is low and due to chlorophyll. These range from extremely clear type I (e.g., Sargasso Sea) to murkier, but still blue, type III (up-welling regions, colder waters). It also includes nine types of coastal water, where absorption and scattering can be quite high and is dominated by dissolved organic water that absorbs yellow light the least. These are labeled 1 through 9 in rank of increasing murkiness and wavelength of peak transmission. Type 9 is essentially yellow. While arbitrary, this system is useful, and if someone tells you that the water at your coral reef site is Jerlov type II, you will have a good idea how it transmits light. For more on the fascinating subject of light in the ocean, Jerlov's *Marine Optics*, published in 1976, is comprehensive and at the same time short and easy to read.

The table below gives the percentage of down-welling irradiance left at 10 meters depth in all five oceanic water types and the clearest coastal type. It is useful for calculations, but also for getting a sense of the wavelength dependence of light transmission in the real world.

Wavelength (nm)	I	Ia	Ib	II	III	1
350	60	53	46	27	10	3.5
375	74	66	58	36	14	5.8
400	80	73	65	42	18	8.1
425	83	76	67	44	20	9.3
450	84	77	70	49	25	13

Wavelength (nm)	I	Ia	Ib	II	III	1
475	83	78	72	54	31	18
500	76	72	67	53	35	23
525	60	58	55	46	33	23
550	53	51	49	42	32	24
575	39	38	37	33	26	20
600	9.0	8.7	8.5	7.5	5.9	4.7
625	4.2	4.1	3.9	3.4	2.6	2.0
650	2.8	2.7	2.6	2.2	1.5	1.1
675	1.3	1.2	1.2	1.0	0.71	0.52
700	0.15	0.15	0.14	0.13	0.12	0.10

Useful Formulas

Black Body Radiation

c speed of light in a vacuum (2.99×10^8 m/s)
h Planck's constant (6.63×10^{-34} Joule seconds)
k Boltzmann's constant (1.38×10^{-23} Joules/Kelvin)
T temperature of black body (Kelvin)

... in wavelength and quantal units
$$L(\lambda) = \frac{2c}{\lambda^4} \frac{1}{e^{\frac{hc}{\lambda kT}} - 1}$$

... in wavelength and energy units
$$L(\lambda) = \frac{2hc^2}{\lambda^5} \frac{1}{e^{\frac{hc}{\lambda kT}} - 1}$$

... in frequency and quantal units
$$L(\nu) = \frac{2\nu^2}{c^2} \frac{1}{e^{\frac{h\nu}{kT}} - 1}$$

... in frequency and energy units
$$L(\nu) = \frac{2h\nu^3}{c^2} \frac{1}{e^{\frac{h\nu}{kT}} - 1}$$

Visibility (For Horizontal Viewing)

d viewing distance
c beam attenuation coefficient
L_o radiance of object at zero distance
$L(d)$ radiance of object at distance d

L_b radiance of background
C_o contrast at zero distance (inherent contrast)
C_{min} minimum contrast detectable by viewer

Radiance of extended object $L_o e^{-cd} + L_b \left(1 - e^{-cd} \right)$

Contrast of extended object $\dfrac{L(d) - L_b}{L_b} = C_o e^{-cd}$

Sighting distance of extended object $\dfrac{\ln\left(\dfrac{|C_o|}{C_{min}} \right)}{c}$

SCATTERING AND REFLECTION

n_1 refractive index of surrounding medium
n_2 refractive index of particle
m refractive index ratio n_2/n_1
λ wavelength of incident light in medium
V volume of particle
θ angle of incidence of light
R reflectance of material
E irradiance striking material
R_{wet} reflectance of wet object

Scattering cross section of small particle (Rayleigh) $\dfrac{24\pi^3 V^2}{\lambda^4} \left(\dfrac{m^2 - 1}{m^2 + 2} \right)^2$

Angle after passing into new medium (Snel's law) $\sin^{-1}\left(\dfrac{\sin\theta}{m} \right)$

Angle of total internal reflection $\sin^{-1}(m)$

Reflected radiance from diffuse opaque surface $\dfrac{RE}{\pi}$

Reflectance of a thick clear slab (parallel polarization)

$$\left(\frac{\cos\theta-\sqrt{m(m-\sin^2\theta)}}{\cos\theta+\sqrt{m(m-\sin^2\theta)}}\right)^2$$

(perpendicular polarization)

$$\left(\frac{\cos\theta-\frac{1}{m}\sqrt{\frac{1}{m}(m-\sin^2\theta)}}{\cos\theta+\frac{1}{m}\sqrt{\frac{1}{m}(m-\sin^2\theta)}}\right)^2$$

Reflectance of a thin clear slab of thickness t $\dfrac{2r(1-\cos\delta)}{1+r^2-2r\cos\delta}$, where

$$\delta=\frac{4\pi n t}{\lambda}\cos\theta,\text{ and}$$

$$r=\left(\frac{m-1}{m+1}\right)^2$$

GEOMETRIC OPTICS AND VISION

i distance from center of lens to focused image
o distance from center of lens to object
f focal length of lens
D diameter of lens
N_o number of photons absorbed by a photoreceptor in one integration
   time

Lens/mirror equation $\dfrac{1}{o}+\dfrac{1}{i}=\dfrac{1}{f}$

Magnification $\dfrac{i}{o}$

F-number $\dfrac{f}{D}$

Width of Airy disk (diffraction spot) on retina $2.44\,f\dfrac{\lambda}{D}$

Minimum contrast threshold (C_{min}) $\dfrac{2.77}{\sqrt{N_0}}$

POLARIZATION

I first Stokes parameter (intensity)

Q second Stokes parameter (linear polarization, vertical vs. horizontal; see text)

U third Stokes parameter (linear polarization, $+45°$ vs. $-45°$; see text)

V fourth Stokes parameter (circular polarization, right-handed vs. left-handed; see text)

I_{max} maximum intensity of transmitted light after passing through a polarizer

I_{min} minimum intensity of transmitted light after passing through a polarizer

I_0 intensity of transmitted light after passing through a vertically oriented polarizer

I_{45} intensity of transmitted light after passing through a $45°$ oriented polarizer

I_{90} intensity of transmitted light after passing through a horizontally oriented polarizer

m refractive index ratio

Degree of polarization $p = \dfrac{\sqrt{Q^2 + U^2}}{I}$, or $p = \dfrac{I_{max} - I_{min}}{I_{max} + I_{min}}$

Angle of polarization (from Stokes parameters)

$$\phi = \tfrac{1}{2}\tan^{-1}\left(\frac{U}{Q}\right) \text{, if } Q \geq 0$$

$$\phi = \tfrac{1}{2}\tan^{-1}\left(\frac{U}{Q}\right) - 90 \text{, if } Q < 0 \text{ and } U < 0$$

$$\phi = \tfrac{1}{2}\tan^{-1}\left(\frac{U}{Q}\right) + 90 \text{, if } Q < 0 \text{ and } U \geq 0$$

Angle of polarization (using polarizer positions)

$$\phi = \tfrac{1}{2}\tan^{-1}\left(\frac{I_{90}+I_0-2I_{45}}{I_{90}-I_0}\right)+90 \text{, if } I_{90} \ge I_0$$

$$\phi = \tfrac{1}{2}\tan^{-1}\left(\frac{I_{90}+I_0-2I_{45}}{I_{90}-I_0}\right)+180 \text{, if } I_0 > I_{90} \text{ and } I_0 > I_{45}$$

$$\phi = \tfrac{1}{2}\tan^{-1}\left(\frac{I_{90}+I_0-2I_{45}}{I_{90}-I_0}\right) \text{, if } I_0 > I_{90} \text{ and } I_0 \le I_{45}$$

Brewster's angle (totally polarized reflection) $\tan^{-1}(m)$

SOLAR ELEVATION (θ)

$$\theta = \sin^{-1}\left(\cos(h)\cos\delta\cos\Phi + \sin\delta\sin\Phi\right)$$

where:

h is the solar hour angle ($=15$ times the number of hours after noon in degrees)

Φ is the geographic latitude in degrees

δ is the solar declination, which (in degrees) is well approximated by:

$$\delta \cong 23.5^\circ \cos\left(\frac{360^\circ}{365}(N+10)\right), \text{ where } N \text{ is the number of days after}$$

January 1.

Equipment and Software Suppliers

GENERAL OPTICS

Ocean Optics — I've bought all my spectrometers and many optical accessories (e.g., fiber optic cables, reflectance standards, integrating spheres) from this small company. Their equipment is (relatively) cheap, portable, and reliable, and their technical support is easy to reach and useful. Their software, however, is not nearly as well designed and can be unreliable. While this is frustrating, over all I've been happy with them. By the way, the "ocean" in their name comes from their location on the beach in Dunedin, Florida. They do not specialize in underwater light measurements and their equipment is NOT waterproof. **Stellarnet**, based literally down the street in Tampa, sells what seems to be the same suite of products with different packaging (I just bet there's an interesting story behind this). They claim their equipment is more rugged than that of Ocean Optics, which would be a big selling point for field biologists, but I have never used any of it.

Edmund Optics — This company, based in Barrington, New Jersey, grew out of Edmund Scientific, which originally specialized in surplus optics and science kits. Now they supply a diverse array of optics, mounts, cameras, microscopy parts, and light sources for prices that are slightly lower than the other major optics suppliers. They also supply certain optics, such as lenses, in different quality grades, which allows you to choose how much you want to spend. They are the first place I look for optical equipment (aside from the spectrometers and fiber optic cables I buy from Ocean Optics).

Interlectric Corporation — This company, based in Warren Pennsylvania, sells a diverse array of lamps, especially fluorescent bulbs. Useful for behavioral experiments that require even illumination with odd spectra. Great prices.

CVI Melles-Griot — An optics supplier like Edmund Optics, but with a larger inventory and somewhat fancier (and more expensive) equipment. The only things I buy from them are filters, because they have a great selection. **Rolyn Optics** and **Esco Products** both also sell a lot of filters, but at a lower price. They are thinner and more fragile, but still good.

Newport/Oriel — Used to be two companies until Newport bought Oriel. More or less a clone of Melles-Griot, but known for their excellent (and expensive) optical tables.

Optronics Laboratories Very nice and superexpensive spectrometers. The best place to go for high-end scanning spectroradiometers. Be prepared to pay at least $50,000.

Newport Glass — Can make large custom filters for you. This is useful, because most standard filters are no bigger than 25 mm in diameter.

Hamamatsu Photonics — The place to go for photomultipliers and low-light cameras. These guys specialize in capturing every photon. You pay for it, but they do deliver.

Bolder Vision Optik — A good source for polarizing filters. The main optical companies also sell polarizers, but these guys have ones that work in the UV and IR. They also sell wave plates and liquid-crystal devices that allow you to change the polarization angle electronically.

Labsphere — While this company sells spectrometers, light sources, and various standards, they are famous for their integrating spheres, some of which are big enough to stand up inside. I don't know what you would do with a six-foot-diameter sphere and god knows what it costs, but—if you need it— it's available.

HobiLabs — This company mostly makes specialized ocean optics instruments, but they make one thing that is indispensable for the aquatic biologist— a cosine corrector that works underwater. Not only is it waterproof, but it gives the correct cosine response in water. A typical cosine corrector won't give you the right answer under water because light behaves differently at the water-plastic interface than at an air-plastic interface.

Electrophysics — Source for infrared imaging equipment. Nice selection of cameras and night vision systems. They used to make cheap UV-transmissive lenses, but don't seem to anymore. Instead try **Universe Kogaku** for relatively cheap lenses for UV photography.

NightSea — Specializes in filter and lights designed to detect fluorescence under water. These are fun to use on a night dive on a coral reef.

Nikon/Canon — I won't step into the Nikon/Canon debate except to say that both are excellent and, for scientific purposes at least, head and shoulders above the rest of the camera companies. Nobody else has the same range of cameras, lenses, and accessories, and both companies are at the top of the field for sensitivity and resolution. The consumer cameras are so good now that they can, in some cases, replace more specialized equipment like intensified and microscope cameras.

Environmental Radiometers and Spectrometers

Li-cor Biosciences — This company makes a wide range of irradiance sensors, mostly for terrestrial work. They are highly portable and reliable, but none of the current models measure spectra. Instead you get integrated measurements. They used to make the LI-1800, a spectroradiometer that was popular for terrestrial work.

International Light Technologies — Make battery-powered and plug-in radiometers for field use. Like Li-cor, though, they all seem to take only integrated measurements.

Yankee Environmental Systems — Among other things, makes high-end stand-alone UV, visible, and IR spectrometers and multiband radiometers. I have never bought them, but they look nice in the catalog.

Unihedron — Make what they call "sky quality meters" that measure the brightness of the nighttime sky (in magnitude per arc second). I think they are primarily purchased by amateur and professional astronomers who want to measure the darkness of a night sky. However, with a few filters and other alterations, they may be useful for studying nocturnal illumination.

Biospherical Instruments — This company makes a number of different instruments that measure radiance and irradiance under water. Their ultraviolet irradiance instruments are the standard in the field. This company will also work with you to build a custom instrument (for the right price).

Satlantic — Like Biospherical, this company makes underwater radiance and irradiance instruments. Some of them incorporate filters that match those in the ocean-observing satellites, allowing you to ground-truth remote sensing data.

WETLabs Inc. — Essentially the only company that makes an instrument that measures the absorption and extinction coefficient of water *in situ*.

This instrument, known as the "ac-9" (because it works at nine wavelengths), is the standard in the field. Fussy to use, but indispensable if you need to know the optical properties of a natural body of water.

SOFTWARE

Sequoia Scientific — Sells Hydrolight, which is probably the most user-friendly radiative transfer program around. Most of the other packages are only for people who could write their own code, anyway. It allows you to model the underwater light field and how it's affected by a slew of parameters including water type, chlorophyll concentration, solar elevation, wind, and cloud cover. When I bought the software ten years ago it cost $10,000. However you also get a T-shirt, lifetime upgrades, and amazing tech support in the form of a personal relationship with the programmer, Curtis Mobley. It's mostly used for ocean optics, but you can use it for anything ranging from clouds to paint to skin, as long as you know the absorption and scattering coefficients and have some idea of the volume scattering function. As Curt will be the first to tell you, though, "garbage in, garbage out."

Remcom — Makes software that allows you to model light propagation through an arbitrary object using the finite difference time domain method. I haven't used it, but the user interface looks better than any other I have seen.

Mieplot — Free Mie scattering software written by Philip Laven. Allows you to change all the relevant parameters and plot the results in various ways including rainbow plots that let you see how scattering by large numbers of a given particle would look in our sky. Simple, clean, and you can't beat the price. Highly recommended.

Stellarium — Another piece of free software that realistically depicts the day and night sky from any earth location at any time from 100,000 BCE to 100,000 CE. The renderings are beautiful and the data quite accurate. Even accounts for the proper motion of stars if you want to see what the Big Dipper will look like in 50,000 years. Also highly recommended.

ImageJ — Free image processing and analysis software originally developed by the National Institutes of Health.

BIBLIOGRAPHY

Aas, E. (1981). The refractive index of phytoplankton. *Univ. Oslo Rep. Ser.* 46.

Ala-Laurila, P., Donner, K., and A. Koskelainen (2004a). Thermal activation and photoactivation of visual pigments. *Biophys. J.* 86, 3653–62.

Ala-Laurila, P., Pahlberg, J., Koskelainen, A. and K. Donner (2004b). On the relation between the photoactivation energy and the absorbance spectrum of visual pigments. *Vision Res.* 44, 2153–58.

Arnold, K. E., Owens, I.P.F., and N. J. Marshall (2002). Fluorescent signaling in parrots. *Science* 295, 92.

Aspelmeyer, M., Böhm, H. R., Gyatso, T., Jennewein, T., Kaltenbaek, R., Lindenthal, M., Molina-Terriza, G., Poppe, A., Resch, K., Taraba, M., Ursin, R., Walther, P., and A. Zeilinger (2003). Long-distance free-space distribution of quantum entanglement. *Science* 301, 621–23.

Balch, W. M., Kilpatrick, K. A., and C. C. Trees (1996). The 1991 Coccolithophore bloom in the Central North Atlantic. 1. Optical properties and factors affecting their distribution. *Limnol. Oceanogr.* 41, 1669–83.

Barr, E. S. (1955). Concerning index of refraction and density. *Am. J. Phys.* 9, 623–24.

Bass M. (1995). *Handbook of Optics.* McGraw-Hill: New York (two edited volumes).

Bell, J. S. (1964). On the Einstein Podolsky Rosen paradox. *Physics* 1, 195–200.

Benedek, G. B. (1971). Theory of the transparency of the eye. *Appl. Opt.* 10, 459–73.

Bernard, G. D., and R. Wehner (1977). Functional similarities between polarization vision and colour vision. *Vision Res.* 17, 1019–28.

Berthier, S. (2000). *Les couleurs des Papillons.* Springer-Verlag: Paris.

Beuthan, J., Minet, O., Helfmann, J., Herrig, M., And G. Mueller (1996). That spatial variation of the refractive index in biological cells. *Phys. Med. Biol.* 41, 369–82.

Bohren, C. F. (1987). *Clouds in a Glass of Beer: Simple Experiments in Atmospheric Physics.* John Wiley and Sons: New York.

———. (1991). *What Light Through Yonder Window Breaks? More Experiments in Atmospheric Physics.* John Wiley and Sons: New York.

Bohren C. F., and E. E. Clothiaux (2006). *Fundamentals of Atmospheric Radiation.* Wiley-VCH: Weinheim, Germany.

Bohren, C. F., and D. R. Huffman (1983). *Absorption and Scattering of Light by Small Particles.* John Wiley & Sons: New York.

Bond, D. S., and F. P. Henderson (1963). *The Conquest of Darkness.* Document number 346297, Defense Documentation Center, Alexandria, VA.

Bone, R. A., and J. T. Landrum (1983). Dichroism of lutein: A possible basis for Haidinger's brushes. *Appl. Opt.* 22, 775–76.

Brenner, M. P., Hilgenfeldt, S., and D. Lohse (2002). Single-bubble sonoluminescence. *Rev. Mod. Phys.* 74, 425–84.

Camara, C. G., Escobar, J. V., Hird, J. R., and S. J. Putterman (2008). Correlation between nanosecond X-Ray flashes and stick-slip friction in peeling tape. *Nature* 455, 1089–92.

Caveney, S., and P. McIntyre (1981). Design of graded-index lenses in the superposition eyes of scarab beetles. *Phil. Trans. Roy. Soc. Lond. B* 294, 589–632.

Chapman, G. (1976). Transparency in organisms. *Experientia* 15, 123–25.

Charney, E., and F. S. Brackett (1961). The spectral dependence of scattering from a spherical alga cell and its implication for the state of organization of the light accepting pigments. *Arch. Biochem. Biophys.* 92, 1–12.

Chiao, C.-C., Cronin, T. W., and D. Osorio (2000). Color signals in natural scenes: characteristics of reflectance spectra and effects of natural illuminants. *J. Opt. Soc. Am.* A 2, 218–24.

Chiou, T.-H., Cronin, T. W., Caldwell, R. L., and N. J. Marshall (2005). Biological polarized light reflectors in stomatopod crustaceans. *SPIE* 5888, 380–88.

Chiou, T-H, Kleinlogel, S., Cronin, T. W., Caldwell, R., Loeffler, B., Siddiqi, A., Goldizen, A., and N. J. Marshall (2008). Circular polarization vision in a stomatopod crustacean. *Curr. Biol.* 18, 429–34.

Chiou, T.-H., Marshall, N. J., Caldwell, R. L., and T. W. Cronin (2011). Changes in light-reflecting properties of signalling appendages alter mate choice behaviour in a stomatopod crustacean *Haptosquilla trispinosa. Mar. Fresh. Behav. Physiol.* 44, 1–11.

Cinzano, P., Falchi, F., and C. D. Elvidge (2001). The first world atlas of the artificial night sky brightness. *Month Not. Roy Astron. Soc.* 328, 689–707.

Collier, T., Arifler, D., Malpica, A., Follen, M., and R. Richards-Kortum (2003). Determination of epithelial tissue scattering coefficient using confocal microscopy. *IEEE J. Sel. Top. Quant. Elec.* 9, 307–13.

Costello, M. J., Johnsen, S., Frame, L., Gilliland, K. O., Metlapally, S., and D. Balasubramanian (2010). Multilamellar spherical particles as potential sources of excess light scattering in human age-related nuclear cataracts. *Exp. Eye Res. 91, 881–89.*

Cronin, T. W., Shashar, N., Caldwell, R. L., Marshall, J., Cheroske, A. G., and T. H. Chiou (2003). Polarization vision and its role in biological signalling. *Integr. Comp. Biol.* 43, 549–58.

Crookes, W. J., Ding, L-L., Hunag, Q. L., Kimbell, J. R., Horwitz, J., and M. J. Mc-

Fall Ngai (2004). Reflectins: The unusual proteins of squid reflective tissues. *Science* 303, 235–38.

Dacke, M., Nilsson, D.-E., Scholtz, C. H., Byrne, M., and E. J. Warrant (2003). Animal behaviour: Insect orientation to polarized moonlight. *Nature* 424, 33.

Dacke, M., Nilsson, D.-E., Warrant, E. J., Blest, A.. D., Land., M. F., and D. C. O'Carroll (1999). Built-in polarizers form part of a compass organ in spiders. *Nature* 401, 470–73.

D'Angelo, G. J., Glasser, A., Wendt, M., Williams, G. A., Osborn, D. A., Gallagher, G. R., Warren, R. J., Miller, K.V., and M. T. Pardue (2008). Visual specialization of an herbivore prey species, the white-tailed deer *Can. J. Zool.* 86: 735–43.

Dehlinger, D., and M. W. Mitchell (2002). Entangled photons, nonlocality, and Bell inequalities in the undergraduate laboratory. *Am. J. Phys.* 70, 903–10.

Delaye, M., and A. Tardieu (1983). Short-range order of crystalline proteins accounts for eye lens transparency. *Nature* 302, 415-–17.

Denton, E. J. (1970). On the organization of reflecting surfaces in some marine animals. *Phil. Trans. R. Soc. Lond. B* 258, 285–313.

Denton, E. J., Gilpin-Brown, J. B., and P. G. Wright (1972). The angular distribution of the light produced by some mesopelagic fish in relation to their camouflage. *Proc. R. Soc. Lond. B* 182, 145–58.

Denton E. J., and M. F. Land (1971). Mechanisms of reflexion in silvery layers of fish and cephalopods. *Proc. Roy. Soc. Lond. A.* 178, 43–61.

Denton, E. J., and J.A.C. Nicol (1965). Studies on reflexion of light from silvery surfaces of fishes, with special reference to the bleak *Alburnus alburnus. J. Mar. Biol. Assoc. UK* 45, 683–703.

Deustch, D. (1998). *The Fabric of Reality: The Science of Parallel Universes and Its Implications.* Penguin: New York.

Douglas, R. H., Partridge, J. C., Dulai, K. S., Hunt, D. M., Mullineaux, C. W., and P. H. Hynninen (1999). Enhanced retinal longwave sensitivity using a chlorophyll-derived photosensitiser in *Malacosteus niger*, a deep-sea dragon fish with far red bioluminescence. *Vision Res.* 39, 2817–32.

Douglas, R. H., Partridge, J.C., Dulai, K., Hunt, D., Mullineaux, C. W., Tauber, A. Y., and Hynninen, P. H. (1998) Dragon fish see using chlorophyll. *Nature* 393, 423–24.

Doyle, W. T. (1985). Scattering approach to Fresnel's equations and Brewster's law. *Am. J. Phys.* 53, 463–68.

Dukes, J. P., van Oort, B.E.H., Stokkan, K., Stockman, A., and G. Jeffery (2003). Seasonal adaptations to LL and DD in the eyes of Arctic reindeer. Society for Neuroscience Abstract Viewer and Itinerary Planner Volume; 2003 Abstract No. 286.4.

Duntley, S.Q. (1952). The visibility of submerged objects. Final Report to Office of Naval Research.

Einstein, A., Podolsky, B., and N. Rosen (1935). Can quantum-mechanical description of physical reality be considered complete? *Phys. Rev.* 47, 777–80.

Engheta, N., Salandrino, A., and A. Al'u (2005). Circuit elements at optical frequencies: Anoinductors, nanocapacitors, and nanoresistors. *Phys. Rev. Lett.* 95, 095504.

Everett, H. (1957). Relative state formulation of quantum mechanics. *Rev. Mod. Phys.* 29, 454–62.

Fasel, A., Muller, P.-A., Suppan, P., and E. Vauthey (1997). Photoluminescence of the African scorpion "Pandinus imperator." *Photochem. Photobiol. Sci.* 39, 96–98.

Feynman, R. P. (1985). *QED: The Strange Theory of Light and Matter.* Princeton University Press: Princeton, NJ.

Firsov, M. L., and V. I. Govardovskii (1990). Dark noise of visual pigments with different absorption maxima. *Sensornye Sistemy* 4, 25–34.

Fox, D. L. (1976). *Animal Biochromes and Structural Colours.* University of California Press: Berkeley.

Frenzel, H., and H. Schultes (1934). Lumineszenz im ultraschall-beschickten Wasser. *Z. Phys. Chem. Abt. B* 27B, 421–24.

Frisch, K. von (1967). *The Dance Language and Orientation of Bees.* The Belknap Press of the Harvard University Press: Cambridge, MA.

Garstang, R. H. (2004). Mount Wilson Observatory: The sad story of light pollution. *Observatory* 124, 14–21.

Giese, A. C. (1959). Comparative physiology: Annual reproductive cycles of marine invertebrates. *Ann. Rev. Physiol.* 21, 547–76.

Goddard, S. M., and R.B.J. Forward (1991). The role of the underwater polarized light pattern in sun compass navigation of the grass shrimp *Palaemonetes vulgaris. J. Comp. Physiol. A* 169, 479–91.

Govardovskii, V. I., Fyhrquist, N., Reuter, T., Kuzmin, D. G., and K. Donner (2000). In search of the visual pigment template. *Vis. Neurosci.* 17, 509–28.

Griswold, M. S., Stark, W. S. (1992). Scotopic spectral sensitivity of phakic and aphakic observers extending into the near ultraviolet. *Vision Res.* 32, 1739–43.

Grooth, B. G. de (1997). Why is the propagation velocity of a photon in a transparent medium reduced? *Am. J. Phys.* 65, 1156–64.

Haddock, S.H.D., and J. F. Case (1995). Not all ctenophores are bioluminescent: Pleurobrachia. *Biol. Bull.* 189:356–62.

——— (1999). Bioluminescence spectra of shallow and deep sea gelatinous zooplankton: Ctenophores, medusae and siphonophores. *Mar. Biol.* 133:571–82.

Haddock, S.H.D., Dunn, C. W., Pugh, P. R., and C. E. Schnitzler (2005). Bioluminescent and red-fluorescent lures in a deep-sea siphonophore. *Science* 309, 263.

Haddock, S.H.D., Moline, M. A., and J. F. Case (2010). Bioluminescence in the sea. *Ann. Rev. Mar. Sci.* 2, 443–93.

Harvey, E. N. (1957). *A History of Luminescence from the Earliest Times Until 1900.* J. H. Furst Co.: Baltimore, MD.

Hawrysyhn, C. W. (1992). Polarization vision in fish. *Am. Sci.* 80, 164–75.

Helfenstein, P., Veverka, J., and J. Hillier (1997). The lunar opposition effect: A test of alternative models. *Icarus* 128, 2–14.

Helfman, G. S. (1986). Fish behaviour by day, night and twilight. pp. 366–-87. In T. J. Pitcher (ed.). *The Behaviour of Teleost Fishes*, Croom Helm, London.

Herring P. J. (1987). Systematic distribution of bioluminescence in living organisms. *J. Biolum. Chemilum.* 1:147–63.

Herring, P. J., and C. Cope (2005). Red bioluminescence in fishes: on the suborbital photophores of *Malacosteus, Pachystomias* and *Aristostomias. Mar. Biol.* 148, 383–94.

Herring, P.J., Widder, E. A., and S.H.D. Haddock (1992). Correlation of bioluminescence emissions with ventral photophores in the mesopelagic squid *Abralia veranyi. Mar. Biol.* 112, 293–98.

Hobson. E. S. (1968). Predatory behavior of some shore fishes in the Gulf of California. *U.S. Bureau of Sport Fisheries and Wildlife Research Report* 73: 1–92.

Hoeppe, G. (2007). *Why the Sky Is Blue: Discovering the Color of Life.* Princeton University Press: Princeton, NJ.

Horváth, G., Kriska, G., Malik, P., and B. Robertson (2009). Polarized light pollution: A new kind of ecological photopollution. *Front. Ecol. Environ.* 7, 317–25.

Horváth, G., and D. Varjú (2004). *Polarized Light in Animal Vision—Polarization Patterns in Nature.* Springer-Verlag: New York.

Hulbert, O. (1953). Explanation of the brightness and color of the sky, particularly the twilight sky. *J. Opt. Soc. Am. A* 43, 113–18.

Isayama, T., Alexeev, D., Makino, C. L., Washington, I., Nakanishi, K., and N. J. Turro (2006). An accessory chromophore in red vision. *Nature* 443, 649–49.

Ivanoff, A., and T. H. Waterman (1958). Factors, mainly depth and wavelength, affecting the degree of underwater light polarization. *J. Mar. Res.* 16, 283–307.

Izumi, M., Sweeney, A. M., Demartini, D., Weaver, J. C., Powers, M. L., Tao, A., Silvas, T. V., Kramer, R. M., Crookes-Goodson, W. J., Mäthger, L. M., Naik, R. R., Hanlon, R. T., and D. E. Morse (2010). Changes in reflectin protein phosphorylation are associated with dynamic iridescence in squid. *J. Roy. Soc. Inter.* 7, 549–60.

Jacquez, J. A., Huss, J., McKeehan, W., Dimitroff, J. M., and H. F. Kuppenheim (1955). Spectral reflectance of human skin in the region 0.7–2.6 um. *J. Appl. Physiol.* 8, 297–99.

Jerlov, N. G. (1976). *Marine Optics.* Elsevier: Amsterdam.

Johnsen, S. (1994). Extraocular sensitivity to polarized light in an echinoderm. *J. Exp. Biol.* 195: 281–91.

———— (2001). Hidden in plain sight: The ecology and physiology of organismal transparency. *Biol. Bull.* 201: 301–38.

———— (2002). Cryptic and conspicuous coloration in the pelagic environment. *Proc. R. Soc. Lond. B* 269, 243–56.

———— (2005).The red and the black: Bioluminescence and the color of animals in the deep sea. *Integr. Comp. Biol.* 45, 234–46.

Johnsen, S., Kelber, A., Warrant, E. J., Sweeney, A. M., Lee, R. H. Jr., and J. Hernández-Andrés (2006). Crepuscular and nocturnal illumination and its effects on color perception by the nocturnal hawkmoth *Deilephila elpenor. J. Exp. Biol.* 209, 789–800.

Johnsen, S., Marshall, N. J., and E. A. Widder (2011). Polarization sensitivity as a contrast enhancer in pelagic predators: Lessons from *in situ* polarization imaging of transparent zooplankton. *Phil. Trans. Roy. Soc. Lond. B* 366, 655–70.

Johnsen, S., Widder, E. A.., and C. D. Mobley (2004). Propagation and perception of bioluminescence: Factors affecting the success of counterillumination as a cryptic strategy. *Biol. Bull.* 207, 1–16.

Kelber, A., Balkenius, A., and E. J. Warrant (2002). Scotopic colour vision in nocturnal hawkmoths. *Nature* 419, 922–25.

Kerker, M. (1969). *The Scattering of Light and Other Electromagnetic Radiation.* Academic Press: New York.

Kidd, R., Ardini, J., and A. Anton (1989). Evolution of the modern photon. *Am. J. Phys.* 57, 27–35.

Kinoshita, S. (2008). *Structural Colors in the Realm of Nature.* World Scientific Publishing Company: Hackensack, NJ.

Kinoshita, S., Yoshioka, S., and J. Miyazaki (2008). Physics of structural colors. *Rep. Prog. Phys.* 71, 1–30.

Kodric-Brown, A. (1989) Dietary carotenoids and male mating success in the guppy: An environmental component to female choice. *Behav. Ecol. Sociobiol.* 25, 393–401.

Können, G. P. (1985). *Polarized Light in Nature.* Cambridge University Press: New York.

Korringa, P. (1947). Relations between the moon and periodicity in the breeding of marine animals. *Ecol. Monogr.* 17, 347.

Kubelka P., and Munk F. (1931). Ein Beitrag Zur Optik der Farbanstriche, *Zeitschrift fur technische Physik* 12, 593–601.

Lamb, W. E. (1995). Anti-photon. *Appl. Phys. B* 60, 77–84.

Land, M. F., and D.-E. Nilsson (2002). *Animal Eyes.* Oxford University Press: New York.

Lawrence, S. J., Lau, E., Steutel, D., Stopar, J. D., Wilcox, B. B., and P. G. Lucey (2003). A new measurement of the absolute spectral reflectances of the moon. *Lunar Planet. Sci.* 34, 1269–70.

Leibowitz, H. W., and D. A Owens (1991). Can normal outdoor activities be carried out during civil twilight? *Appl. Opt.* 30, 3501–503.

Levi, L. (1974). Blackbody temperature for threshold visibility. *Appl. Opt.* 13, 221.

Liu, Z., Yanagi, K., Suenaga, K., Kataura, H., and S. Iijima (2007). Imaging the dynamic behaviour of individual retinal chromophores confined inside carbon nanotubes. *Nat. Nanotechnol.* 2, 422–25.

Lohse, D., Schmitz, B., and M. Versluis (2001). Snapping shrimp make flashing bubbles. *Nature* 413, 477–78.

Longcore, T., and C. Rich (2004). Ecological light pollution. *Front. Ecol. Environ.* 2, 191–98.

Lynch, D. K., and W. Livingston (1995). *Color and Light in Nature.* Cambridge University Press: Cambridge, UK.

Lythgoe, J. N. (1979). *The Ecology of Vision.* Clarendon Press: Oxford.

Marcikic, I., Riedmatten, H. de, Tittel, W., Zbinden, H., Legre, M., and N. Gisin (2004). Distribution of time-bin entangled qubits over 50 km of optical fiber. *Phys. Rev. Lett.* 93, 180502 (4 pp).

Marshall, B. R., and R. C. Smith (1990). Raman scattering and in-water ocean optical properties. *Appl. Opt.* 29, 71–84.

Marshall, N. J., Cronin, T W., Shashar, N., and M. F. Land (1999). Behavioural evidence for polarisation vision in stomatopods reveals a potential channel for communication. *Curr. Biol.* 9, 755–58.

Mathger, L. M., Barbosa, A , Miner, S., and R. T. Hanlon (2006). Color blindness and contrast perception in cuttlefish (*Sepia officinalis*) determined by a visual sensorimotor assay. *Vision Res.* 46, 1746–53.

Mattila, K. (1980). Synthetic spectrum of the integrated starlight between 3000 and 10000 Å. Part 1. Method of calculation and results. *Astron. Astrophys. Ser.* 39, S53–S65.

Matz, M. V., Fradkov, A. F., Labas, Y. A., Savitsky, A. P., Zaraisky, A. G., Markelov, M. L., and S. A. Lukyanov (1999). Fluorescent proteins from nonbioluminescent Anthozoa species. *Nature Biotech.* 17, 969–73.

Mazel, C. H., Cronin, T. W., Caldwell, R. L., and N. J. Marshall (2004). Fluorescent enhancement of signaling in a mantis shrimp. *Science* 303, 51.

Mazel, C. H., and E. Fuchs (2003). Contribution of fluorescence to the spectral signature and perceived color of corals. *Limnol Oceanogr.* 48, 390–401.

McFarland, W. N., Wahl, C., Suchanek, T., and McAlary, F. (1999). The behavior of animals around twilight with emphasis on coral reef communities. In *Adaptive Mechanisms in the Ecology of Vision* (eds. S. N. Archer, M.B.A. Djamgoz, E. R. Loew, J. C. Partridge, and S. Vallerga), pp. 583–628. Kluwer Academic: Boston.

McNamara, G., Gupta, A., Reynaert, J., Coates, T. D., and C. Boswell (2006). Spectral imaging microscopy web sites and data. *Cytometry A.* 69, 863–71.

Meinel, A., and M. Meinel (1983). *Sunsets, Twilights, and Evening Skies.* Cambridge University Press: New York.

Mermin, N. D. (1985). Is the moon there when nobody looks? Reality and the quantum theory. *Phys. Today* 38(4), 38–47.

Mertens, L. E. (1970) *In-Water Photography: Theory and Practice.* John Wiley & Sons: New York.

Michiels, N. K., Anthes, N., Hart, N. S., Herler, J., Meixner, A. J., Schleifenbaum, F., Schulte, G., Siebeck, U. E., Sprenger, D., and M. F. Wucherer (2008). Red fluorescence in reef fish: a novel signaling mechanism? *BMC Ecol* 8, 16 pages.

Minnaert, M. (1954). *The Nature of Light and Color in the Open Air.* Dover: New York.

Mobley, C. D. (1994). *Light and Water Radiative Transfer in Natural Waters.* Academic Press: San Diego, CA.

Moore, M. V., Pierce, S. M., Walks, H. M., Kvalvik, S. K., and J. D. Lim (2000). Urban light pollution alters the diel vertical migration of *Daphnia. Verh. Internat. Verein. Limnol.* 27, 1–4.

Moran, N. A., And T. Jarvik (2010). Lateral Transfer of Genes from Fungi Underlies Carotenoid Production in Aphids. *Science* 328, 624–27.

Nassau, K. (1983). *The Physics and Chemistry of Color: The Fifteen Causes of Color.* John Wiley & Sons Inc: New York.

Neville, A. C., and S. Caveney (1969). Scarabaeid beetle exocuticle as an optical analogue of cholesteric liquid crystals, *Biol. Rev.* 44, 531–62.

Novales Flamarique, I., Hawryshyn, C. W., and F. I. Harosi (1998). Double cone internal reflection as a basis for polarized light detection. *J. Opt. Soc. Am. A* 15, 349–58.

Nyholm, S. V., M. J. McFall-Ngai. (2004). The Winnowing: Establishing the Squid–*Vibrio* Symbiosis. *Nat. Rev. Microbiol.* 2, 632–42.

Odom, B., Hanneke, D., D'Urso, B., and G. Gabrielse (2006). New measurement of the electron magnetic moment using a one-electron quantum cyclotron. *Phys. Rev. Lett.* 97.

Ohmiya, Y., Kojima, S., Nakamura, M., and H. Niwa (2005). Bioluminescence in the limpet-like snail, *Latia nerotoides. Bull. Chem. Soc. Jpn.* 78, 1197–205.

Palmer, J. M. (1995). The measurement of transmission, absorption, emission, and reflection. In *Handbook of Optics II* (eds. M. Bass, E. W. Van Strylan, D. R. Williams, and W. L. Wolfe), pp. 25.1–25.25. McGraw-Hill Inc.: New York.

Parker, A. R., Hegedus, Z., and R. A. Watts (1998). Solar-absorber antireflector on the eye of an Eocene fly. *Proc. R. Soc. Lond. B* 265, 811–15.

Parker, A. R., McPhedran, R. C., McKenzie, D. R., Botten, L. C., and N.-A. P. Nicorovici (2001). Aphrodite's iridescence. *Nature* 409, 36–37.

Patek, S. N., and R. L. Caldwell (2005). Extreme impact and cavitation forces of a biological hammer: Strike forces of the peacock mantis shrimp *Odontodactylus scyllarus. J. Exp. Biol.* 208, 3655–64.

Pendry, J. B., and D. R. Smith (2004). Reversing light with negative refraction. *Phys. Today* 57, 37–43.

Petzold, T. J., and R. W. Austin (1976). *Submerged reflectance.* Scripps Institution of Oceanography Technical Report. Scripps Institution of Oceanography: UC San Diego.

Planck, M. (1901). On the law of distribution of energy in the normal spectrum. *Annal. Phys.* 4, 1–6

Pomozi, I, Horváth, G., Wehner, R. (2001). How the clear-sky angle of polarization pattern continues underneath clouds: Full-sky measurements and implications for animal orientation. *J. Exp. Biol.* 204, 2933–42.

Prum, R. O., Torres, R. H., Williamson, S., and J. Dyck (1998). Coherent light scattering by blue bird feather barbs. *Nature* 396, 28–29.

——— (1999). Two-dimensional Fourier analysis of the spongy medullary keratin of structurally coloured feather barbs. *Proc. R. Soc. Lond. B* 266, 13–22.

Purves, D., and R. B. Lotto (2003). *Why We See What We Do: An Empirical Theory of Vision.* Sinauer: Sunderland, MA

Quinby-Hunt, M. S., and A. J. Hunt (1988). Effects of structure on scattering from marine organisms: Rayleigh-Deybe and Mie predictions. *SPIE Ocean Optics IX* 925, 288–95.

RCA Electro-Optics Handbook (1974). Burle Industries: Lancaster, PA.

Reif, F. (1965). *Fundamentals of Statistical and Thermal Physics.* McGraw-Hill: New York.

Rickel, S., and A. Genin (2005). Twilight transitions in coral reef fish: The input of light-induced changes in foraging behavior. *Anim. Behav.* 70, 133–44.

Roberts, N. W., Chiou, T.-H., Marshall, N. J., and T. W. Cronin (2009). A biological quarter-wave retarder with excellent achromaticity in the visible wavelength region. *Nature Photonics* 3, 641–44.

Rossel, S., and R. Wehner (1982). The bee's map of the e-vector pattern in the sky. *Proc. Nat. Acad. Sci.* 79, 4451–55.

Rozenberg, G. V. (1966). *Twilight: A Study in Atmospheric Optics.* Plenum Press: New York.

Salart, D. Baas, A.., Branciard, C., Gisin, N., and H. Zbindin (2008). Testing the speed of "spooky actions at a distance." *Nature* 454, 861--64.

Schechner, Y. Y., and N. Karpel (2005). Recovery of underwater visibility and structure by polarization analysis. *IEEE J. Ocean Eng.* 30, 570–87.

Schechner, Y. Y., Narasimhan, S. G., and S. K. Nayar (2003). Polarization-based vision through haze. *Appl. Opt.* 42, 511–25.

Schwind, R. (1989) A variety of insects are attracted to water by reflected polarized light. *Naturwissenschaften* 76, 377–78.

Seapy, R. R., and R. E. Young (1986). Concealment in epipelagic pterotracheid heteropods (Gastropoda) and cranchiid squids (Cephalopoda). *J. Zool. Lond.* 210, 137–47.

Segelstein, D. (1981).*The Complex Refractive Index of Water*. M.S. Thesis, University of Missouri–Kansas City.

Seidou, M., Sugahara, M., Uchiyama, H., Hiraki, K., Hamanaka, T., Michinomae, M., Yoshihara, K., and Y. Kito (1990). On the three visual pigments in the retina of the firefly squid, *Watasenia scintillans. J. Comp. Physiol. A* 166, 769–73.

Shashar, N., Hagan, R., Boal, J. G., and R. T. Hanlon (2000). Cuttlefish use polarization sensitivity in predation on silvery fish. *Vision Res.* 40, 71–75.

Shashar, N., Hanlon, R. T., and A. D. Petz (1998). Polarization vision helps detect transparent prey. *Nature* 393, 222–23.

Shashar, N., Rutledge, P. S., and T. W. Cronin (1996). Polarization vision in cuttlefish—a concealed communication channel? *J. Exp. Biol.* 199, 2077–84.

Shevell, S. K. (2003).*The Science of Color.* Elsevier: New York (edited volume).

Shurcliff, W. A. (1962). *Polarized Light: Production and Use.* Harvard University Press: Cambridge, MA.

Siefken, S.K.C. (2010). "Seasonal changes in the *Tapetum lucidum* as an adaptation to winter darkness in reindeer." Master's thesis. University of Tromso, Norway.

Sivak, J. G., and T. Mandelman (1982). Chromatic dispersion of the ocular media. *Vision Res.* 22, 997–1003.

Smith, R. C., and K. S. Baker (1981). Optical properties of the clearest natural waters (200–800 nm). *Appl. Opt.* 20, 177–84.

Smith, R. L. (1970). The velocities of light. *Am. J. Phys.* 38, 978–84.

Smith, V. C., and J. Pokorny (2003). Color matching and discrimination. In *The Science of Color* (ed., S. K. Shevell), pp. 103–48. Elsevier: New York.

Soffer, B. H., and D. K. Lynch (1999). Some paradoxes, errors, and resolutions concerning the spectral optimization of human vision. *Am. J. Phys.* 11, 946–53.

Sommerfeld, A. (1954). *Optics: Lectures on Theoretical Physics, Volume IV*. Academic Press: New York.

Speiser, D. I., and S. Johnsen (2008). Scallops visually respond to the presence and speed of virtual particles. *J. Exp. Biol.* 211, 2066–70.

Stavenga, D. G., Foletti, S., Palasantzas, G., and K. Arikawa (2006). Light on the moth-eye corneal nipple array of butterflies. *Proc. R. Soc. Lond. B* 273, 661–67.

Stavenga, D. G., Smits, R. P., and B. J. Hoenders (1993). Simple exponential functions describing the absorbance bands of visual pigment spectra. *Vision Res.* 33, 1011–17.

Stavenga, D. G., Stowe, S., Siebke, K., Zeil, J., and K. Arikawa (2004). Butterfly wing colours: Scale beads make white pierid wings brighter. *Proc. R. Soc. Lond. B* 271, 1577–84.

Stenner, M. D., Gauthier, D. J., and M. A. Neifeld (2003). The speed of information in a "fast-light" optical medium. *Nature* 425, 695–98.

Sutton, T. T., and T. L. Hopkins (1996). Trophic ecology of the stomiid (Pisces: Stomiidae) fish assemblage of the eastern Gulf of Mexico: Strategies, selectivity and impact of a top mesopelagic predator group. *Mar. Biol.* 127, 179–92.

Sweeney, A. M., Des Marais, D. L., Ban, Y. A., and S. Johnsen (2007). Evolution of graded refractive index in squid lenses. *J. Roy. Soc. Interface* 4: 685–98.

Sweeney, A. M., Jiggins, C., and S. Johnsen (2003). Polarized light as a butterfly mating signal. *Nature* 423: 31–32.

Sweeting, L. M. (1990). Light your Candy. *ChemMatters* 8(3), 10–12.

Taylor, F. W. (2005). *Elementary Climate Physics*, Oxford University Press: Oxford, UK.

Treibitz, T., Schechner, Y. Y., and H. Singh (2008). Flat refractive geometry. *Proc. IEEE CVPR.*

Twomey, S. A., Bohren, C. F., and J. L. Mergenthaler (1986). Reflectance and albedo differences between wet and dry surfaces. *Appl. Opt.* 25, 431–37.

Vaezy, S., and J. I. Clark (1994). Quantitative analysis of the microstructure of the human cornea and sclera using 2-D Fourier methods. *J. Microscop.* 175: 93–99.

Van de Hulst, H. C. (1957). *Light Scattering by Small Particles.* Dover: New York.

Van Dover, C. L., Reynold, G. T., Chave, A. D., and J. A. Tyson (1996). Light at deep-sea hydrothermal vents. *Geophys. Res. Lett.* 23, 2049–52.

Van Dover, C. L., Szuts, E. Z., Chamberlain, S. C., and J. R. Cann (1989). A novel eye in "eyeless" shrimp from hydrothermal vents of the mid-Atlantic ridge. *Nature* 337, 458–60.

Vigneron, J. P., Pasteels, J. M., Windsor, D. M., Vértesy, Z., Rassart, M., Seldrum, T., Dumont, J., Deparis, O., Lousse, V., Biró, L. P., Ertz, D., V. Welch (2007). Switchable reflector in the Panamanian tortoise beetle *Charidotella egregia* (Chrysomelidae: Cassidinae). *Phys. Rev. E* 76, 031907.

Voss, K. J., and H. Zhang (2006). Bidirectional reflectance of dry and submerged Labsphere Spectralon plaque. *Appl. Opt.* 45, 7924–27.

Vukusic, P., and J. R. Sambles (2003). Photonic structures in biology. *Nature* 424, 852–55.

Vukusic, P., Sambles, J. R., and C. R. Lawrence (2000). Colour mixing in wing scales of a butterfly. *Nature* 404, 457.

Wagner, H.-J., Douglas, R. H., Frank,, T. M., Roberts, N. W., and J. C. Partridge (2009). A novel vertebrate eye using both refractive and reflective optics. *Curr. Biol.* 19, 1–7.

Walker, J. (2006). *The Flying Circus of Physics*. Wiley: New York.

Walton, A. J. (1977).Triboluminescence. *Advance Phys.* 26, 887–948.

Warrant, E. J. (2006). Invertebrate vision in dim light. In *Invertebrate Vision* (ed. E. J. Warrant and D.-E. Nilsson), Cambridge: Cambridge University Press, pp. 83–126.

Warrant, E. J., and Nilsson, D.-E. (1998). Absorption of white light in photoreceptors. *Vision Res.*, 38, 195–207.

Washington, I., Zhou, J., Jockusch, S., Turro, N. J., Nakanishia, K., and J. R. Sparrow (2007). Chlorophyll derivatives as visual pigments for super vision in the red. *Photochem. Photobiol. Sci.* 6, 775–79.

Waterman, T. H. (1981). Polarization sensitivity. In *Handbook of Sensory Physiology* (ed., H. Autrum), vol. 7/6B, New York: Springer, pp. 281–469.

——— (2006). Reviving a neglected celestial underwater polarization compass for aquatic animals. *Biol. Rev.* 81, 111–15.

Wehner, R., and G. D. Bernard (1993). Photoreceptor twist: A solution to the false-color problem. *Proc. Nat. Acad. Sci.* 90, 4132–35.

Wehner, R., and Labhart, T. (2006). Polarization vision. In *Invertebrate Vision* (eds., E. J. Warrant and D. E. Nilsson), pp. 291–348. Cambridge University Press: Cambridge, UK.

White, S. N., Chave, A. D., and G. T. Reynolds (2002). Investigation of ambient light emission at hydrothermal vents. *J. Geophys. Res.* 107, 1–13.

Widder, E. A. (1998). A predatory use of counterillumination by the squaloid shark, *Isistius brasiliensis. Env. Biol. Fish* 53, 267–73.

———. (2010). Bioluminescence in the ocean: Origins of biological, chemical, and ecological diversity. *Science* 328, 704–708.

Wild, F. J., Jones, A. C., and A. W. Tudhope (2000). Investigation of luminescent banding in solid coral: The contribution of fluorescence. *Coral Reefs* 19, 132–40.

Wynberg, H., Meijer, E. W., Hummelen, J. C., Dekkers, H.P.J.M., Schippers, P. H., and A. D. Carlson (1980). Circular polarization observed in bioluminescence. *Nature* 286, 641–42.

Wyszecki, G., and Stiles, W. S. (1982). *Color Science: Concepts and Methods, Quantitative Data and Formulae*. New York: John Wiley and Sons.

Yahel, R., Yahel, G., Berman, T., Jaffe, J., and A. Genin (2005). Diel pattern with abrupt crepuscular changes of zooplankton over a coral reef. *Limnol. Oceaongr.* 50, 930–44.

Yang, H., Xie, S., Li, H., and Z. Lu (2007). Determination of human skin optical properties *in vivo* from reflectance spectroscopic measurements. *Chin. Opt. Lett.* 5, 181–83.

INDEX